COOPERATIVE APPRENTICESHIPS

HOW TO ORDER THIS BOOK

BY PHONE: 800-233-9936 or 717-291-5609, 8AM–5PM Eastern Time

BY FAX: 717-295-4538

BY MAIL: Order Department
Technomic Publishing Company, Inc.
851 New Holland Avenue, Box 3535
Lancaster, PA 17604, U.S.A.

BY CREDIT CARD: American Express, VISA, MasterCard

BY WWW SITE: http://www.techpub.com

Cooperative Apprenticeships

A School-to-Work Handbook

JEFFREY A. CANTOR, Ph.D.

Director of Technical Education

Virginia Community College System, Richmond, Virginia

LANCASTER · BASEL

Cooperative Apprenticeships
a**TECHNOMIC**® publication

Published in the Western Hemisphere by
Technomic Publishing Company, Inc.
851 New Holland Avenue, Box 3535
Lancaster, Pennsylvania 17604 U.S.A.

Distributed in the Rest of the World by
Technomic Publishing AG
Missionsstrasse 44
CH-4055 Basel, Switzerland

Printed in the United States of America
10 9 8 7 6 5 4 3 2 1

Main entry under title:
Cooperative Apprenticeships: A School-to-Work Handbook

A Technomic Publishing Company book
Bibliography: p.
Includes index p. 285

Library of Congress Catalog Card No. 96-61085
ISBN No. 1-56676-471-8

A project of this magnitude requires motivation and inspiration.
I am fortunate to be surrounded by a loving family
that constantly provides me with both motivation
and inspiration, as well as love.
To my wife Ruth, daughter Julie, and sons
David and Adam, I dedicate this book.

Table of Contents

Preface

APPRENTICESHIP as structured workplace learning is witnessing renewed interest as a process for preparing American workers. This results from studies and publications such as the William T. Grant Foundation Commission's report, "The Forgotten Half" (1988) and the "Workforce 2000" report, among others. The findings of these studies call for development of cooperative work strategies and experiential learning to facilitate development of critical thinking on the part of young learners and workers and movements towards expanding apprenticeship as a "paradigm for learning" (Berryman in Rosenbaum et al., 1992). As offered by President Bill Clinton, in the report "The Forgotten Half," workers under the age of twenty-five who have only a high school diploma are earning 28 percent less than they were fifteen years ago. Clinton calls for the development of an American system of apprenticeship, blending vocational and academic education in high school. The goal is to provide students with meaningful work experience linked to continued education and training after graduation. Legislation aimed at facilitating school-to-work initiatives continues to underwrite this movement.

A major thrust of school-to-work initiatives, cooperative apprenticeship programs are increasingly popular because they also serve as a catalyst for linking secondary and vocational-technical schools and business to promote cooperation for worker preparation and local economic development. Educators and administrators in secondary and vocational schools, technical institutes, and community colleges are, in fact, working cooperatively with trainers in business and industry and are moving rapidly towards developing cooperative apprenticeship programs. The focus of this work is to provide guidance for structured workforce training program development and delivery through cooperative apprenticeship based upon proven and workable practices in education and industry. This book is written as a handbook to provide guidance to instructors and administrators responsible for developing, delivering, and administering apprenticeship training cooperatively with local businesses and/or government organizations. The book is a resource for program operation and a "how-to" guide for effective communications between education and business and industry.

REFERENCE

Rosenbaum, J. E., Stein, D., Hamilton, S., Hamilton, M., Berryman, S. and Kazis, R. (1992). *Youth Apprenticeship in America: Guidelines for Building an Effective System.* Washington, D.C.: William T. Grant Foundation, Commission on Work.

CHAPTER 1

What Is Cooperative Apprenticeship?

THIS book is a guide for designing and developing apprenticeship education and training programs for today's youth and tomorrow's skilled workers. With changing American worker demographics, workplace technologies, and increasing world economic competition, workforce preparation becomes a critical issue. Industry, labor, education, and government leaders are concerned with those issues relating to education and training strategies that will improve the skills of America's current and future workers. We all share in the problem and must share in the solution. A key to the solution is shared responsibility for worker preparation based upon industry-derived training standards. Apprenticeship offered cooperatively with vocational-technical schools and community colleges meets this purpose.

In recent years, questions have been raised about the readiness of today's workforce in terms of meeting the competitive demands of an international economy. Employers frequently report that too many young aspiring workers come to the workplace with serious deficiencies, leaving them poorly equipped to perform on the job. The problems raised are often attributed to our system of public education. To be competitive, American business and industry must lead the way in providing standards-based education and training to meet their demands for a skilled workforce. American business is meeting and continues to meet this goal. In fact, conservative estimates of between $300 and $400 million are spent annually in the construction and manufacturing industries alone on apprenticeship training.

One particular strategy—registered apprenticeship—has captured the interest of many educators, business leaders, and policymakers.

WHAT IS APPRENTICESHIP?

Apprenticeship is a proven training strategy that combines on-the-job training with related classroom instruction. It has historically involved a formal arrangement of employers, employer associations, labor unions, and state governments. Apprenticeship is an industry-based basic or initial training process. By definition,

> Apprenticeship is characterized by a contractual employment relationship in which the firm or sponsor promises to make available a broad and structured practical and theoretical training of an established length and/or scope in a recognized occupational skill category. Apprenticeship is a work-study training scheme in which part of the training occurs on the job and part occurs off the job in a classroom or workshop setting. (Glover, 1980)

Apprenticeship, in simplest terms, is training in occupations that require a wide and diverse range of skills and knowledge, as well as maturity and independence of judgement. It involves planned, day-by-day training on the job, and experience under proper supervision, combined with related technical instruction (U.S. DOL, 1991).

The term *cooperative apprenticeship,* as used in this book, is a form of structured workplace training in which (1) an employer, employer group, or an industry, labor union, or other community-based organization *joins with* a vocational-technical school and/or community college to provide formal instruction in which the structured work-based experience is an integral part of the instruction; (2) an apprentice agrees to work for the employer for a specified period of time; (3) the employer agrees through a registered agreement to provide structured and formal training in a specific field or trade over a defined period of time; and (4) the employer provides continued journeyman-level employment after the training is successfully completed.

Experience has demonstrated that apprenticeship is a highly effective strategy for preparing people for work. While much of U.S. apprenticeship is in the construction and manufacturing industries, there is significant potential to develop apprenticeship programs in a variety of other industries. What benefits are derived to employers, workers, and society as a whole from sponsoring registered apprenticeship?

Questions are asked about the adaptability of our workers to maintain their current skills levels while learning new and emerging technologies. A training system in which workers can continually update their skills and learn from the masters preceding them in the trade or occupation is needed. Apprenticeship offered cooperatively in conjunction with vocational-technical schools and community colleges provides this avenue.

WHY CONSIDER APPRENTICESHIP?

An understanding of cooperative apprenticeship is important for vocational-technical school and community college educators and administrators as, increasingly, employers and public policymakers are looking to school-based professionals to help promote and support apprenticeship. Today, apprenticeship is more than an industry-based on-the-job training process, which is the historical definition of the process; it now is a cooperative educational and training venture between educational institutions, employers, industry associations, and the community-at-large. This book will treat this new approach to workforce preparation—cooperative apprenticeship.

And what about tomorrow's workforce? American values respective to occupational choice for the next generation are coming under scrutiny as many school-age youngsters are opting out of the skilled trades and occupations—oftentimes, without the proper guidance and information. Most technical trades and occupations are requiring increasing cognitive abilities and levels of education. Many presently require the equivalent of two years of postsecondary education—hence, an Associate degree. It is imperative that we continue to devise instructional strategies that provide youth with a positive view of craft work and potential occupations and with opportunities for acquiring higher education. Our workforce is not now, nor will it ever be, a totally white-collar or professional workforce. Again, cooperatively sponsored apprenticeship offers these opportunities.

Today's learner is perhaps more perceptive and inquisitive than ever before. Today's trainee needs to know why certain knowledge and skills are needed before mastery can occur. Hence, an effective training system must provide the opportunity to motivate the trainee and provide realistic conditions under which learning takes place. Training must be realistic, delivered when and where needed, and provide for immediate feedback about level of trainee mastery. Again, cooperatively sponsored apprenticeship best meets these requirements.

The 1988 Grant Commission report entitled, "Youth Apprenticeship in America" (Rosenbaum et al., 1992) reminds us that workers who are age twenty-five and under with only a high school diploma are earning 28 percent less today than they were fifteen years ago; and high school dropouts are even worse off, earning 42 percent less today than fifteen years ago. It is all too well recognized that our present educational system falls short of being a "world-class" system. Too many youth leave secondary schools without career direction or preparation for work. Too many youth fail to graduate with diplomas at all.

Our workforce has changed in make-up and composition. The supply of workers is different today than it was in past decades and centuries. Employers must be prepared to recruit from varying populations and be ready to work with and train these newly assimilated Americans. All American workers must be prepared to interface with peoples of different backgrounds, beliefs, customs, and languages. This mandates that preparation for cultural awareness be a component of workforce preparation. Cooperative apprenticeship provides for opportunities to learn in dynamic educational, work, and social environments.

As more and more technical careers require an increased related education and, in many cases, postsecondary education including the Associate degree, apprenticeship today also involves cooperation between educational professionals in secondary and vocational schools and two-year community-technical colleges and the student who wants to learn a skilled occupation or trade. Apprenticeship permits business and industry to train tomorrow's skilled workforce by today's journeymen and -women. It permits the firm to invest in human resource development and get a ready return on its investment.

Further, the trainee is more motivated and career-minded inasmuch as he/she can learn and earn. The Federal Committee on Apprenticeship indicates that, today, over 40,000 industries and companies offer registered apprenticeship training to approximately 340,000 apprentices representing all walks of life. Many of these activities are in the construction and manufacturing industries in about 34,000 different training programs.

Apprenticeship is good for economic development within a community. Firms are able to hire workers when and where needed and train them to industry standards. Trainees are able to earn and learn, unemployment is reduced, and firms have the needed talent to grow and prosper. Today, more than ever before, American business faces fierce competition from abroad. Employers must be able to meet the competition with a flexible, cost-effective, and rapidly trained workforce. Only cooperative apprenticeship permits an employer to train an employee to industry-based and -controlled standards, when and where needed, and have that trainee able to provide immediate services back to the employer while the trainee continues to learn and earn towards full journeyman status. The trainee earns a livable wage during training, and the employer has a useful worker during training.

These are the essential ingredients to promote local economic development; therefore, what cooperative apprenticeship affords employers, trainees, and society in general is a system of training and education that motivates both learner and employer, provides for long-term career development with a significant investment on both parties' parts, and gives workers actual training for existing workforce requirements. However, apprenticeship definitely has benefits to offer to other industries, including service, retail, and public sector industries. Expanded apprenticeship can provide high-quality training for tomorrow's workers.

A BRIEF HISTORY AND OVERVIEW OF APPRENTICESHIP

To appreciate the beauty of apprenticeship as a training tool to meet the needs of American education and business, a look at its history is in order.

I begin with a brief, but comprehensive, overview of traditional apprenticeship as a process for occupational training.

Apprenticeship is not a new educational concept. It is one of the oldest forms of training in which the skills and knowledge associated with the skilled occupations are passed from a skilled worker to a learner. Evidence of apprenticeship training has been found in ancient Egyptian tombs. The Babylonian Code of Hammurabi provided a written account of a system of apprenticeship existing in 2100 B.C. During the 13th and 14th centuries, expert craftsmen, such as silversmiths, weavers, coach makers, and blacksmiths, formed trade groups called guilds to maintain the highest possible standards of quality and workmanship in their individual trades. One of the main duties of the historical guild master was to train apprentices to carry on the skills of the trades. A youth, usually at the age of about sixteen, was assigned to a master craftsman whose trade was to be learned. This apprentice not only worked for and learned from the master, but actually lived in the same home as part of the family during the apprenticeship, which lasted several years.

As a form of structured training, U.S. apprenticeship dates back to colonial days. In the employer's shop, the apprentice was taught the skills of the trade, spending hours working under the careful supervision of the master. Work was checked every step of the way for skill and accuracy. In addition to learning the "secrets of the trade," the apprentice also learned to be industrious, reliable, and proud of good work. Such a skilled and honest craftsman was assured a position of honor and prosperity in the community.

Since the Middle Ages, skills have been passed on through a master-apprentice system in which an apprentice was indentured to the master for a specified period of years. The apprentice usually received food, shelter, and clothing in return for the work the apprentice preformed while indentured. For instance, craft worker apprenticeship in the early U.S. colonial leather industry is reflected in the Indenture Agreement of Gould Brown:

> North Kingston, April the 7th 1792. We the subscribers this day have mutually agreed that I Gould Brown, am to work with Mr. Benjamin Greene the term of twenty four months, for the sum of three pounds lawful silver money to me in hand paid at the expiration of said time; and the said Benjamin is to allow the said Gould Brown the Privilege of Tanning and Curring Six Calves Skins and two large sizes only tan'd; and is to find him two pair of thick Double Sould shoes, and as many frocks and trousers to ware as he needs in the tan-yard to work, and to Board him the said Gould Brown and Wash his Clothes the same time. Further, I the said Gould Brown, Do agree to Bring with me One Sett of Shoemakers tools for to work with, and Mr. Benjamin Greene agrees to let him have another Sett to Bring away with When his time is Expired. (U.S. Department of Labor, 1982, p. 6)

This system was in widespread use until the industrial revolution created the need for more structure. With the expansion of industry after the industrial revolution, changes took place in the apprenticeship system to conform to the requirements of the machine age. The domestic apprenticeship disappeared. No longer did the apprentice live with the master. The term *indenture* acquired negative meanings during the period around the American Revolution, when indentured servants were a major source of the new workers in the colonies. At the hands of unscrupulous people, however, indentured servants were treated poorly, which gave the word *indenture* its negative connotations.

In the public sector, U.S. Navy apprenticeship training programs are perhaps the oldest continually operating apprenticeships in the United States. The Navy's first "apprentice boy" was hired in the Washington Navy Yard in 1810. The first formal apprentice school in the U.S. Navy opened at Mare Island Navy Yard in 1858 when that yard was under the administration of David Farragut. It remains one of the most vital forms of training for today's military and civilian defense worker, especially in high-technology areas.

In the U.S. construction industry where apprenticeship has its roots, documentation illustrates that, in 1832, a "house carpenter" was indentured in New Bedford, Massachusetts to his master until 1837. The indenture states that John Slocum "doth by these Presents bind Lyman Slocum (age 16), his son, a minor . . . to Thomas Remington . . . to learn the art, trade, or mystery of a House-Carpenter" (U.S. DOL, 1982, p. 8). Apprenticeship in the United States continued as an unregulated system until 1911, when Wisconsin passed the country's first apprenticeship law. With safeguards for both the apprentice and the employer, Wisconsin's law became a model both for other states and for the federal government in developing their own systems.

Registered apprenticeship programs have historically been operated by private sector industry, including employers and labor/management sponsors. These program sponsors pay the majority of all training costs, including wages on a progressive schedule to their apprentices. The content of the training program is determined by industry needs, thereby producing workers with skills that are in demand.

ORGANIZED LABOR

Labor unions have long sponsored apprenticeship training programs through their locals nationwide. In the 1920s, national employer and labor organizations, as well as educators and others, lobbied for a national system of apprenticeship. The construction industry was at the forefront of this movement. The post–World War I construction boom, coupled with curtailed immigration, created a need for worker skill training. As a result, The Federal Committee on Apprenticeship, composed of representatives of government agencies, was appointed by the Secretary of Labor to serve as a national policy-recommending body on apprenticeship in the United States. Organized labor spends about $6 billion a year on apprenticeship in the construction industries alone. Herein, there are about 34,000 different programs training several hundred thousand apprentices yearly.

FEDERAL AND STATE LAWS AND POLICIES

The National Apprenticeship Act of 1937 (P.L. 308) was passed and became the legal basis for federal apprenticeship policy (more commonly referred to as the Fitzgerald Act). This Act was designed:

> to promote the furtherance of labor standards of apprenticeship . . . to extend the application of such standards by encouraging the inclusion thereof in contracts of apprenticeship, to bring together employers and labor for the formulation of programs of apprenticeship, to cooperate with State agencies in the formulation of standards of apprenticeship.

TEXTBOX 1.1

National Apprenticeship Act

(50 Stat. 664; 29 U.S.C. 50)

To enable the Department of Labor to formulate and promote the furtherance of labor standards necessary to safeguard the welfare of apprentices and to cooperate with the State in the promotion of such standard.

Be it enacted by the Senate and House of Representatives of the United States of America in Congress assembled, That the Secretary of Labor is hereby authorized and directed to formulate and promote the furtherance of labor standards necessary to safeguard the welfare of apprentices, to extend the application of such standards by encouraging the inclusion thereof in contracts of apprenticeship, to cooperate with State agencies engaged in the formulation and promotion of standards of apprenticeship, and to cooperate with the Department of Interior in accordance with section 6 of the Act of February 23, 1917 (39 Stat. 932), as amended by Executive Order Numbered 6166, June 10, 1933, issued pursuant to an Act of June 30, 1932 (47 Stat. 414), as amended.

Sec. 2. The Secretary of Labor may publish information relating to existing and proposed labor standards of apprenticeship, and may appoint national advisory committees to serve without compensation. Such committees shall include representatives of employers, representatives of labor, educators, and officers of other executive departments, with the consent of the head of any such department.

Sec 3. On and after the effective date of this Act the National Youth Administration shall be relieved of direct responsibility for the promotion of labor standards of apprenticeship as heretofore conducted through the division of apprentice training and shall transfer all records and papers relating to such activities to the custody of the Department of Labor. The Secretary of Labor is authorized to appoint such employees as he may from time to time find necessary for the administration of this Act, with regard to existing laws applicable to the appointment and compensation of employees of the United States: ***Provided, however,*** That he may appoint persons now employed in division of apprentice training of the National Youth Administration upon certification by the Civil Service Commission of their qualifications after nonassembled examinations.

Sec. 4. This Act shall take effect on July 1, 1937, or as soon thereafter as it shall be approved.

Approved, August 16, 1937.

One of the Bureau of Apprenticeship and Training's responsibilities is to register apprentices and issue certificates to apprentices who successfully complete approved programs. For instance, national apprenticeship policy as set forth in Title 29 CFR Part 29 sets out labor standards, policies and procedures relating to the registration, cancellation and de-registration of apprenticeship programs and of apprenticeship agreements by the Bureau of Apprenticeship and Training (BAT), and the recognition of a state apprenticeship council or agency (SAC) as the appropriate agency for registering local apprenticeship programs. (See Appendix A for the U.S. Department of Labor's, "Apprenticeship Programs: Labor Standards for Registration.") Title 29 CFR Part 30 sets forth policies and procedures to promote equality of opportunity in apprenticeship programs registered with the U.S. Department of Labor and in state apprenticeship programs registered with recognized state apprenticeship agencies. These policies and procedures apply to the recruitment and selection of apprentices and to all conditions of employment and training during apprenticeship. There are currently 300,000 registered apprentices—a number that has been relatively constant over the last decade—preparing for over 800 skilled occupations (and many more nontraditional occupations). These regulations also recognize a state's authority.

State apprenticeship laws were adopted to help develop a state's skilled workforce and to help protect those entering the trades. By ensuring a level of uniformity to the training that apprentices receive, it also provides an indirect protective measure to the public who utilizes the structures, products, and services that apprentices and graduates build, make, and provide.

Such laws established three basic procedures to distinguish apprenticeable trades:

(1) There must be a written agreement, an indenture, between the apprentice, the sponsor, and the state. This agreement specifies plainly the length of training, the related school requirements, an outline of the skills of the trade to be learned, and the wages the apprentice will receive.

(2) Work assignments from the employer must allow the apprentice to gain a comprehensive knowledge of the trade.
(3) At the end of the apprenticeship, the graduate must show competency in all the skills of the trade.

LINKING VOCATIONAL EDUCATION AND APPRENTICESHIP

Historically, vocational education and apprenticeship have only coexisted peacefully. In 1911, the Wisconsin legislature established the state's vocational school system to provide the related classroom instruction to apprentices. The Smith-Hughes (1917) and George-Barden (1946) Acts provided for partial reimbursement from federal funds of instructor salaries in states with approved vocational plans. The Carl D. Perkins Vocational & Applied Technology Act Amendments of 1990 provided for a framework for reshaping postsecondary vocational education. The Perkins Vocational Education Act also provides funding for some portions of apprenticeship training. The degree to which this funding is used and the processes used for such linkage vary from state to state. Perkins's monies are used to fund programs in both secondary schools and community colleges.

Current legislative efforts have recently been made to foster better relationships between business and the community. U.S. federal involvement in apprenticeship training is currently rediscovered in the Perkins Vocational Education Act, Job Training Partnership, and Youth Apprenticeship Acts. Specifically, the new Perkins Amendment is aimed at improving collaborative efforts between the secondary school, community college, and industry.

At the time of the writing of this book, President Clinton had just recently signed the Goals 2000 bill (1994), which established the National Skill Standards Board. This board will be charged with endorsing skill standards for broad occupational areas to be developed by consortia of business and industry (American Vocational Association, 1994). Section #3 of the School to Work Opportunities Act of 1994 (P.L. 103-239) stipulates:

> to build on and advance a range of promising school-to-work activities, such as tech-prep education, career academies, school-to-apprenticeship programs, cooperative education, youth apprenticeship, school sponsored enterprises, [and] business-education compacts . . . that can be developed into programs funded under this Act (108STAT.571). (p. 88)

AN AGE-OLD TRADITION CONTINUES TO MEET TODAY'S NEEDS

Yes, historically, the system of apprenticeship has proved to be an effective method for the acquisition of needed workplace skills. It has survived through the ages and is still widely used today. As Van Erden (1991) writes, "Apprenticeship programs depend on the cooperation of private sector organizations that control jobs" (p. 32).

The purpose of this historical perspective, therefore, is to illustrate how apprenticeship has evolved into structured workforce training and how it can be and is a very effective training tool and process for providing quality training for entry-level workers dealing cooperatively with educational institutions. With this historical framework, it is easier to understand why apprenticeship has remained a viable process for training and why it is a wonderful method and process for cooperatively educating and training American youth.

A PRIMER: HOW COOPERATIVE APPRENTICESHIP WORKS

Modern cooperative apprenticeship is a unique training system through which learners acquire state-of-the art "high-tech" trade and craft skills and knowledge in a combination of on-the-job training supervised by highly skilled workers at a job site and related and coordinated vocational-technical school and community college classroom instruction. This instruction can be provided throughout the term of the apprenticeship, from high school to community college, and teaches apprentices the theoretical aspects of their trade through courses such as blueprint reading, mathematics, and sciences. Modern technological disciplines require highly cognitive abilities and complex thinking on the part of workers, which can only be achieved by industry training programs working in combination with higher education. In fact, some schools and colleges have specific programs developed in cooperation with business in which the apprentice earns credit towards a degree for the apprenticeship work. Some states require that employers pay their apprentices for *both* time worked and time spent in required classroom instruction, recognizing the equally important weight of both aspects of apprenticeship training.

In practice, the related theoretical information required to perform the job is learned in classes either after work or during a part of the day set aside for classroom work. For instance, in related instruction classes, numerical control apprentices learn the mathematics that they must know, as well as the quality control processes. Print reading and drawing, computer-assisted design, safety, chemistry, and other sciences related to the work are also learned. Apprentices also gain insights into the economic world, including industrial history, management, and labor relations. This training design provides for mastery of all of the practical and theoretical skills necessary for a chosen occupation, as is experienced in the maritime industry's apprenticeship programs.

TEXTBOX 1.2

The Maritime Industries Cooperative Apprenticeship

The U.S. maritime shipbuilding and repair industries include both U.S. naval shipyards and private sector shipbuilding and repair (overhaul) firms. Collectively, these industries have traditionally been strategically critical to our national defense and our economic competitiveness. Therein, attention to maintaining a trained and competent workforce, even in times of peace, is essential, and cooperative apprenticeship has proven to work for the firms and government organizations that comprise these industries.

Today, mostly all naval shipyards operate some form of apprentice training. Likewise, many, if not all, commercial firms see apprenticeship as a primary means for training their technical and managerial workforces. The U.S. Navy's shipyards (NSYs) each operate apprenticeship training programs for their new technical workers. These programs provide training in approximately forty-three different trade areas, depending on the needs of the shipyard. Despite recent modernization and automation in the industry, effective and safe ship repairs remain dependent upon the knowledge and abilities of large numbers of highly skilled craftspeople. Apprentice-trained journeymen are a majority of each of the shipyard workforces. There are no assembly lines in ship repair. Craftspeople work independently at nonrepetitive tasks and must apply analytical abilities and initiative, as well as judgement, to accomplish their work. Good reading skills are required for interpretation of job orders, technical manuals, and specifications; mathematics supports a variety of work-related calculations.

The apprenticeship programs are important to the shipyards, not only as a source of skilled workmen, but also because apprentice-trained journeymen provide the pool from which supervisors and managers are drawn. Some 87% of the naval shipyard superintendents (heads of the major shop and trade groups) are former apprentices. This system of internal advancement provides shipyards with dedicated upper-level management skilled in ship construction and repair. Herein lies the principal reason for sponsoring higher educational opportunities for these workers, through coordination with local community colleges and other institutions of higher education.

The Navy firmly believes that NSY trade workers should earn a college degree. To these ends, each of the NSYs have worked with their local colleges to make this as feasible as possible for the apprentice.

Not unlike their federal counterparts, private sector yards conduct similar programs using community colleges to provide related instruction and college degree credit.

A SYSTEM BASED ON STANDARDS

One of the most significant benefits of using apprenticeship is the fact that apprenticeship training is built upon industry-derived occupational standards. As will be discussed in Chapter 5, standards for apprenticeship programs are administered by a state department of labor, bureau of apprenticeship and training. Each such bureau jointly reviews classroom training standards with state boards of vocational-technical and/or community college and adult education, which supervise and conduct the related classroom instruction. These bureaus help to ensure that standards are met by the sponsors of apprenticeship (employers). Through a cooperative agreement with the U.S. Department of Labor, field staff from the Federal Bureau of Apprenticeship and Training also work within the states to ensure the smooth functioning of the state's apprentice system and helps ensure that the terms of the apprenticeship are upheld so that the apprentice will receive the necessary training and schooling. The indenture that each apprentice works under is signed, not only by the apprentice and the employer, but also by the state director of the bureau of apprenticeship standards. These bureaus also work closely with state and local apprenticeship committees (often called joint apprenticeship committees, or JACs) for all the major trades and with labor unions, employers, and employer associations to ensure that the state apprenticeship system is the best that it can be.

TOWARDS A MODEL FOR COOPERATIVE APPRENTICESHIP

Again, the term *cooperative apprenticeship* as used in this book is a form of structured workplace training in which (1) an employer, employer group, or an industry, labor union, or other community-based organization *joins with* a vocational-technical school and/or community college to provide formal instruction in which the structured work-based experience is an integral part of the instruction; (2) an apprentice agrees to work for the employer for a specified period of time; (3) the employer agrees to provide structured and formal training in a specific field or trade over a defined period of time; and (4) the employer provides continued journeyman-level employment after the training is successfully completed (see Figure 1.1).

While registered apprenticeship training is usually initiated by the employer, the goals and objectives of apprenticeship from the business training perspective are consistent with the goals and objectives of our public schools and community-technical colleges. Therefore, it is appropriate for vocational educators and administrators to suggest the development of apprenticeships to business and industry to solve their workforce problems as well. For instance, let's look at the *general goals* of apprenticeship:

(1) To develop and ensure a supply of trained, skilled, and knowledgeable workers and supervisors for the operations
(2) To increase worker productivity and overall skills levels and versatility
(3) To lessen the need for supervision of employees by developing initiative, pride of craftsmanship, and speed and accuracy in work

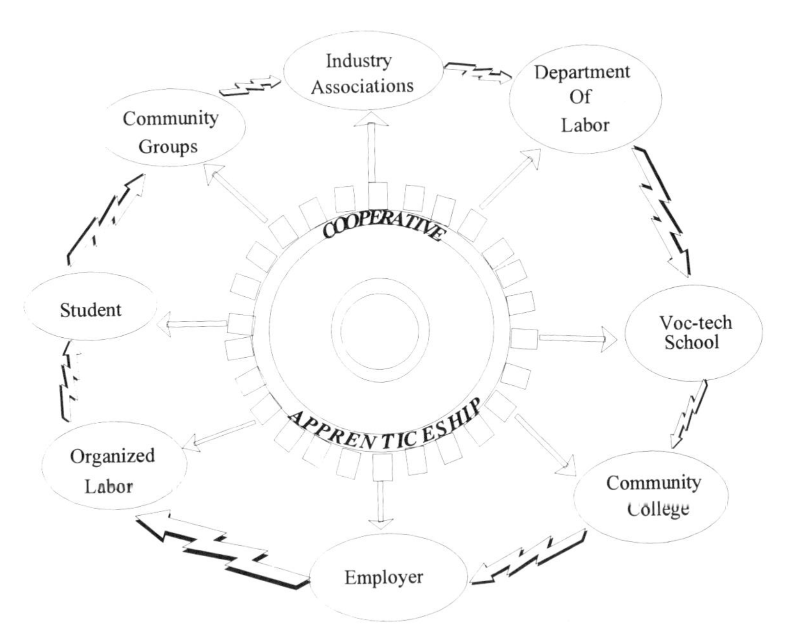

Figure 1.1 Cooperative apprenticeship.

(4) To continue to attract a constant flow of capable workers into the craft or trade

The U.S. Department of Labor's Federal Committee on Apprenticeship (1992) defines apprenticeship as a training strategy with *eight essential components*:

- Apprenticeship is sponsored by employers and others who can actually hire and train individuals in the workplace, and it combines hands-on training on the job with related theoretical instruction.
- Workplace and industry needs dictate key details of apprenticeship programs—training content, length of training, and actual employment settings.
- Apprenticeship has a specific legal status and is regulated by federal and state laws and regulations.
- Apprenticeship leads to formal, official credentials—a certificate of completion and journeyman status.
- Apprenticeship generally requires a significant investment of time and money on the part of employers or other sponsors.
- Apprenticeship provides wages to apprentices during training according to predefined wage scales.

- Apprentices learn by working directly under master workers in their occupations.
- Apprenticeship involves both written agreements and implicit expectations. Written agreements specify the roles and responsibilities of each party; implicit expectations include the right of program sponsors to employ the apprentice, recouping their sizable investment in training, and the right of apprentices to obtain such employment.

Therefore, for cooperative apprenticeship to work to meet these goals and objectives, vocational-technical schools and/or community colleges, local business and industry, and industry groups, as well as government and organized labor and other interested constituencies, must *cooperate*. This book suggests a model system for cooperative apprenticeship. Let's look at what should be each of the roles of the requisite players.

THE ROLE OF THE VOCATIONAL-TECHNICAL SCHOOL

A significant portion of related apprenticeship instruction is conducted in public school facilities. Typically, vocational schools provide related instruction, both classroom and laboratory. This has always been viewed by both industry and public policymakers as beneficial. Even where private industry facilities are used, part of the instructor's salary is often funded by vocational education monies from state and federal sources. Vocational-technical schools do even more than provide quality-related instruction. In Wisconsin, for example, the vocational-technical college system provides administrative support for apprentice on-the-job placement. In Connecticut, the vocational-technical school system oversees the apprentice-related curriculum and instructional development standards. In most all state systems, there is potential for student identification, screening, and placement. There is also opportunity for articulation of related instruction with the community college system to provide an upward educational track for the apprentice to continue an education.

THE COMMUNITY COLLEGE'S ROLE

The community college is ideally positioned to make cooperative apprenticeship work. In fact, I even suggest that it is the community college that should be the primary focal point for cooperative apprenticeship. This institution has the technical expertise and community outreach resources to bring together all of the principal players.

From a business perspective, many community colleges have joined with one or more manufacturers to operate cooperative apprenticeship programs in automotive, maritime, and electronics industries, among others. The National Association of College Automotive Teachers (NACAT) estimates that there are about 500 automotive technology apprenticeship programs operating throughout the United States and Canada. Student recruitment, screening and placement, on-the-job instructor training, curriculum design and maintenance, records keeping, student counseling, and program and apprentice evaluation are all jobs done by community colleges in support of cooperative apprenticeship. The following chapters will describe and illustrate these efforts nationwide.

TEXTBOX 1.3

Automotive Industry Programs

The automotive industry has successfully adopted cooperative apprenticeships as a training process. The car companies believe that, through attaining an Associate degree, a technician will have a desirable general and technical education and will have mastered vital communications, business management, and diagnostic problem-solving skills—all requisites for dealing with customers and changing technology. To train through apprenticeship, local auto dealers, their trade associations, vehicle manufacturers, and independent repair shops work cooperatively with schools and community colleges.

Beginning in the early 1970s, the National Automobile Dealers Association (NADA) recognized a need for a more technically skilled mechanic to meet the ever-increasing automotive technology. Today, automotive manufacturers, including General Motors, Ford Motor Company, Chrysler, American Honda, Toyota, and Nissan Motor Company have such programs.

Manufacturer-oriented curricula are developed in cooperation with vocational school and community college faculty nationwide. The automotive company helps to recruit students; some provide financial assistance in the form of scholarships for tuition, basic tools, and reimbursement for mechanic national certification. Manufacturers commonly donate equipment, training aids, and materials to participating schools. Manufacturers also provide training and development programs for instructors during the summer break periods.

The cooperative apprenticeship idea is a new twist on cooperative education and work-based learning programs, yet it bears a strong resemblance to traditional apprenticeship. In both, there is a system of formal structured training, the student earns a realistic wage, the employer makes a long-term commitment to the student, and the training is for a defined period of time.

ROLE OF ORGANIZED LABOR

American apprenticeship has typically involved industry working cooperatively with organized labor. American apprenticeship today is largely an extension of collective bargaining (MacKenzie, 1983). This is because apprenticeship satisfies some very important union objectives, the most important being the ability to control craft competence and productivity. Secondly, from a collective bargaining perspective, issues include job security and corporate/union investments in job skills development. Union participation in apprenticeship is also a form of union security. By controlling the kind of training, degree of specificity, and numbers of trainees, unions maximize their ability to attract and maintain organized contracts and keep their members versatile and adaptable to work changes. It also serves to prevent substitution of low-wage workers for journeyman members (International Union of Operating Engineers, 1975).

TEXTBOX 1.4

Organized Labor's Programs

"Those who have the proper skills training will be ready to meet the age of technology," according to the International Brotherhood of Electrical Workers (IBEW) President Jack Barry. The IBEW turned 104 years old in 1995. Its 769,000 members are training as journeymen through apprenticeship. The IBEW apprenticeship is a model for the skilled trades.

In the fall of 1977, IBEW Local #3, Flushing, New York ventured forward with an idea promoted by one of its most dynamic leaders, Harry Van Arsdale: to have new apprentices in the construction electricians division of the Local complete an Associate degree in labor studies through Empire State College (NY) as a required part of their apprenticeship training. This decision to add such a requirement and the accompanying effort to implement the educational innovation were unparalleled in the history of organized labor and apprenticeship training.

WHY SUCH A REQUIREMENT?

First, Local #3 and its employer—contractors represented by the National Electrical Contractor's Association (NECA)—view a commitment to education as a well-established tradition in the labor movement. Additionally, the philosophical commitment to the apprenticeship program and to higher education was so strong at that time that the union leadership established the degree program as mandatory in order to serve both the union's and industry's needs. Second, the local leadership viewed the need to raise the educational level of working people, thus improving their social status and self-esteem—and that of all workers as well. They foresaw payoffs for organized labor by impacting positively on the qualifications and characteristics of the bargaining unit, thus, they believed, strategically positioning labor for better union wage packages in collective bargaining.

Next, Local #3's leadership believes that a union concerned about their own public image has a stake in educating both the public and their members about the labor movement (eighteen credits of the Associate degree program for apprentices is in labor studies). Better educated apprentices make better union members—sympathetic to the cause, an argument touted for the program. In all, this college program is part of a total package that includes technical skills training, a job in the trade, membership in the union, and career mobility.

THE CURRICULUM AND INSTRUCTIONAL ISSUES

Local #3's program was designed to provide college credit for on-the-job apprenticeship training, as well as to provide the related trade theory classes held in the evening. The curriculum was adopted from the National Joint Apprenticeship Training Council (NJATC) standards.

As a five-year-long program, apprentices are able to earn forty-seven semester hours towards an Associate degree at participating colleges. Local #3 runs a four-year, state-registered program. After the fourth year, IBEW apprentices are no longer registered with the state, but they are under the obligation/jurisdiction of the union for another eighteen months until they become full journeymen (five and a half years).

ROLE OF THE GOVERNMENT

The Federal Committee on Apprenticeship, as authorized under the Fitzgerald Act, advises the Secretary of Labor on promotion of apprenticeship. Its membership is composed of representatives of labor, business management, and the public. The U.S. Department of Labor's Bureau of Apprenticeship and Training (BAT) is responsible for implementation of the Fitzgerald Act.

At the state level, state apprenticeship councils (SACs) currently exist in twenty-seven states, the Virgin Islands, Guam, Puerto Rico, and the District of Columbia. SACs oversee more than 75 percent of the nation's registered apprenticeship programs (see Figure 1.2, p. 18).

SOME NONTRADITIONAL PLAYERS

Apprenticeship program participation has also included other nontraditional organizations, many serving nontraditional populations (women, Native Americans, prisoners in correctional institutions, and the handicapped). These population groups will be discussed as well in the following chapters.

AN OVERVIEW OF THE BOOK

CHAPTER 2: COOPERATIVE APPRENTICESHIP: BUILDING SCHOOL-TO-WORK COLLABORATIVES

The aim of the book is to describe a process for delivering cooperative workplace training through linkages of business and industry and vocational-technical schools and community colleges through apprenticeship.

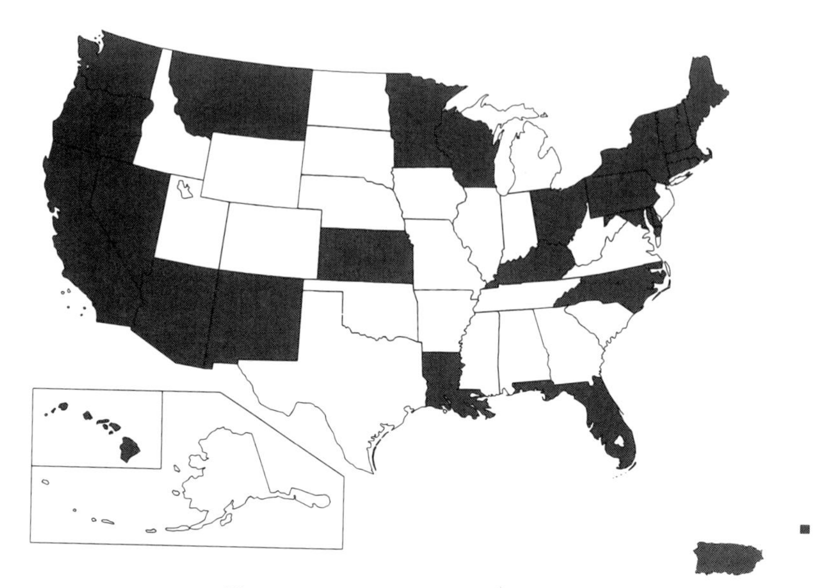

Figure 1.2 States with apprenticeship agencies.

Chapter 2 presents a model for cooperative apprenticeship. First, success for cooperative apprenticeship rests with good communications, which means developing an appreciation for how each party (school and/or college and business and/or government agency) operates. To accomplish this, both educational institutions and businesses must understand the roles and responsibilities of each other—and how these are to be carried out.

The chapter will describe how educational professionals (both vocational-technical school and community college) and business principals can get together to discuss apprenticeship—and where appropriate, with state departments of labor, economic development consortia, and other quasi-governmental entities—to accomplish planning, designing, and delivering cooperative apprenticeship. Successful communication strategies and processes will be described. Case histories of successful model programs will be highlighted throughout the book.

CHAPTER 3: FROM DOING A NEEDS ASSESSMENT TO DECIDING WHEN APPRENTICESHIP IS APPROPRIATE

Chapter 3 is a how-to guide for conducting a needs assessment. Training program development costs money—and in times of tight budgets, business managers are wary of initiating new undertakings without appropriate justification; therefore, a solid needs assessment is an imperative. Determining apprenticeship as appropriate requires an understanding of the factors that contribute to making it successful as an instructional tool, including (1) identification of an occupation or technical profession that is well enough defined in terms of performance standards, (2) identification of firms or businesses capable of providing structured work experiences with accompanying mentoring by master employees, and (3) identifying potential student-workers desiring a formal learn-and-earn situation. The chapter will illustrate how vocational-technical schools and community colleges can work with and aid a local business to conduct a needs assessment to determine if and when apprenticeship is appropriate for the firm or business or, for that matter, whether the educational institution should initiate training for a particular kind of workforce. Chapter 3 will provide guidance to accomplish these tasks. Suggestions and guidance for school and college staffs on marketing apprenticeship will also be provided.

CHAPTER 4: PREAPPRENTICESHIP AND YOUTH APPRENTICESHIP PROGRAMS: A SCHOOL-TO-WORK CONCEPT

The work-based learning concept on which apprenticeship is based is receiving new attention from government and business as a possible solution to keeping kids in school, providing career guidance and direction, and giving them incentives to learn a viable career. Preapprenticeship and youth apprenticeship programs are being viewed as a means to achieving solutions to the productivity, adaptability, and competitiveness problems facing U.S. business and industry. Vocational-technical educators, administrators, employers, and community leaders are working together to build a new generation of preapprenticeship and youth apprenticeship, as well as school-to-work programs.

As will be described in Chapter 4, the School-to-Work Opportunities Act (SWOTA) has "seeded" a number of demonstration programs nationwide, with exemplary sites found in Massachusetts (Boston's Project ProTech) Pennsylvania (Lycoming County) and Maine (youth apprenticeships). The Manpower Development Research Corporation analyzed the practices and experiences of one dozen such programs (Pauly, 1994). This chapter on preapprenticeship will discuss processes for transition from youth to full apprenticeship. These programs focus on students beginning in the ninth or tenth grade to gain their interests before they become disengaged from high school.

CHAPTER 5: DESIGNING AND DEVELOPING APPRENTICESHIP PROGRAMS—PART 1

The focus of Chapter 5 is upon the instructional systems process for designing and developing cooperative apprenticeship programs. This chapter begins with a description of the process for conducting a job/task analysis upon which to develop an apprenticeship program. Employers need a fundamental understanding of how this process is carried out. The process includes identification of appropriate and existing performance standards for the trade or occupation as an alternative for the actual development of same. Key information on the current national vocational education standards projects is provided as a resource. Methods for translation of performance standards into training objectives is also provided.

A key ingredient for program development is the joint apprenticeship committee. It is essential to understand the operation of JACs and how programs can be conducted successfully and cooperatively with these industry/government groups.

Another key ingredient to understand is how to develop agreements to link the employer, trainee, educational institution, and union and/or JAC. This aspect of program operation is covered in depth with ample sample agreements to illustrate the concept. Apprenticeship selection policy guidelines, probationary periods, and employer/union/college advisory committee development and interface are also discussed. Policy guidance on working hours, conditions, supervision, terms of training and employment, and mutual roles and responsibilities is included.

Part of Chapter 5 focuses on highlighting existing apprenticeship programs (in those programs that have a historical underpinning), such as The International Association of Firefighters; International Brotherhood of Electrical Workers; International Union of Operating Engineers; and commercial firms such as Toyota Motors, American Honda, Ford, General Motors, and Chrysler Corp. These programs can serve as resources for development of similar programs by others using this book. State apprenticeship programs such as in Wisconsin, California, and Florida are also highlighted.

CHAPTER 6: DESIGNING AND DEVELOPING APPRENTICESHIPS—PART 2: HOW COOPERATIVE APPRENTICESHIP PROGRAMS ARE DEVELOPED

Chapter 6 addresses the development of programs, program documentation, and coordinating of program delivery. Roles of each of the organizational players are discussed and illustrated. Emphasis is placed on indenturing practices.

Other topics included in Chapter 6 are the elements of successful integration of classroom and on-the-job instruction. The use of measurable performance objectives for linking on-the-job and classroom instruction are highlighted therein.

CHAPTER 7: RECRUITING, SCREENING, AND SELECTING APPRENTICES AND EMPLOYERS

Success in apprenticeship rests with an understanding of how to identify trainees (students) for such programs. Chapter 7 focuses on how to recruit trainees and how the school can aid the firm in recruitment, screening, and selection (and testing) of trainees. Development of checklists and forms for recording progress are discussed. Instructor development, sharing of firm/school employees, meetings with trainees to discuss problems, and progress are all topics of concern. Successful strategies for interview protocols, screening/testing processes, and the like used by some apprenticeship programs such as IBEW's Local #3 NY and California's firefighter JAC programs are discussed.

CHAPTER 8: RELATED CLASSROOM EDUCATION AND EDUCATIONAL SUPPORT

Effective classroom instruction is another key to successful apprenticeship program operations. Firm training personnel need to understand how to cooperate with schools and community colleges to ensure successful classroom instruction. Such personnel need to understand how schools and community colleges operate. Chapter 8 provides this information. It highlights the factors that make classroom instruction successful. The chapter also focuses on the processes necessary for community-technical colleges to grant college degree credit for the total apprenticeship, including the on-the-job portions thereof.

A major portion of the chapter is devoted to administration of related technical education through the school or 2-year college. The procurement of related services is also discussed.

Other important topics include general education and the apprentice, literacy remediation, bilingual education situations, and training the job-site supervisor.

CHAPTER 9: APPRENTICE TESTING AND EVALUATION AND CERTIFICATION

Another key aspect of successful cooperative apprenticeship is testing of the apprentice on the job. The design of appropriate evaluation instruments is presented therein. Development of competency checklists to complement performance objectives for on-the-job use is presented. Supervisor use of these evaluative tools on the job is highlighted. Integration of classroom testing into the apprenticeship is included. The concept of certification of the apprentice and its interrelationship to national organizations is included. National Occupational Competency Testing Institute (NOCTI) program guidelines are discussed.

Lessons learned in exemplary programs are the focus of Chapter 9. There are many fine apprenticeship programs operating nationally, and the highlights of these programs warrant inclusion into a book of this kind. Additionally, there are a number of fine relationships existing between schools and/or community colleges and apprenticeship programs in business and industry. Apprenticeship organizations will be listed and described in this chapter as well.

CHAPTER 10: RESOURCES AND ORGANIZATIONS

Part of operating an apprenticeship program is keeping in touch with others who have like interests and problems. Chapter 10 discusses and describes national, regional, and local organizations that sponsor apprenticeship programs, regular meetings, and lobbying activities. The chapter will also discuss resources available to assist in program development and operation.

REFERENCES

American Vocational Association. (1994, May). *Legislative Update*, p. 1.

Apprenticeship: The Answer for America's Future. (1995). A National Conference. Washington, D.C., October.

Cantor, J. A. (1989). Exemplary practices linking economic development and job training. Presentation to the Educational Management Division, Eastern Educational Research Association, Savannah, GA, February.

Federal Committee on Apprenticeship. (1992). *Youth Training and Education in America: The Role of Apprenticeship.* Washington, D.C.: Author.

Frank H. Cassel and Associates Inc. (1976). *Study in Vocational Education Involvement with Apprenticeship Programs in Illinois, Executive Summary*, Chicago, IL: Frank Cassel and Associates, p. 1025.

Glover, R. W. (1980). Apprenticeship in the United States: Implications for Vocational Education Research and Development. Occasional Paper No. 66. Columbus, OH: The National Center for Research in Vocational Education.

Harford Community College. (1982, April). "The Maryland Association of Electrical Contractors Apprenticeship Joins with HCC's Programs," *Harford Gazette.*

International Union of Operating Engineers. (1975). *Dual Enrollment as an Operating Engineer Apprentice and an Associate Degree Candidate.* Final Report, Washington, D.C.: International Union of Operating Engineers.

Johnston, W. B., and Packer, A. H. (1987). *Workforce 2000: Work and Workers for the 21st Century.* Indianapolis, IN: Hudson Institute.

MacKenzie, J. R. (1983). *Education and Training in Labor Unions, 1982 Report.* Washington, D.C.: Department of Health, Education and Welfare.

Parnell, D. (1990). "The Tech-Prep/Associate Degree Program," *Journal of Studies in Technical Careers*, XII(4):301–305.

Pauly, E. (1994). "Home-Grown Lessons: There's Much to Learn from Existing School-to-Work Programs," *Vocational Education Journal,* 16:69–71.

Regan, W. (1989). "Higher Vocational Goals Eyed by Administration," *Education Week* (March 29):15.

Reubens, B. G. (1980). *Apprenticeship in Foreign Countries*. R&D Monograph no. 77, Washington, D.C.: Department of Labor, Employment and Training Administration.

Rosenbaum, J. E., Stein, D., Hamiton, S., Hamiton, M., Berryman, S., and Kazis, R. (1992). *Youth Apprenticeship in America: Guidelines for Building an Effective System.* Washington, D.C.: William T. Grant Foundation, Commission on Work.

Scott, R. W. (1991, February). "Making the Case for Tech Prep," *Vocational Education Journal*, 22–23 and 63.

U.S. Department of Labor, Bureau of Apprenticeship and Training. (1991, Revised). *National Apprenticeship Program.* Washington, D.C.: Author.

U.S. Department of Labor, Bureau of Apprenticeship and Training. (1982, Revised). *Apprenticeship: Past and Present.* Washington, D.C.: Author.

Van Erden, J. (1991, October). "Linking Past and Present, Students and Jobs," *Vocational Education Journal*, 30–33.

Worthington, R. M. *Public Sector Opportunities and Linkages with Vocational Education.* Washington, D.C.: Office of Vocational and Adult Education, Department of Education, pp. 1–23.

CHAPTER 2

Cooperative Apprenticeship: Building School-to-Work Collaboratives

THIS chapter describes the cooperative apprenticeship model upon which this book is predicated. I begin with effective communications for cooperative apprenticeship. My experience suggests that at the heart of a good cooperative apprenticeship arrangement is communications; therefore, these skills are essential to the success of a cooperative apprenticeship program. I will also discuss the necessary elements—those people and organizations whose participation is of paramount importance to the ultimate success of cooperative apprenticeships.

Cooperative apprenticeship involves four systems that traditionally operate at cross-purposes, but that are critical to effective work-based instruction. These four systems are (1) the employer system, (2) the organized labor (union) system, (3) the economic system, and (4) the vocational-technical school and community-technical college education system (Cantor, 1990). Of course, the student is the focus of the apprenticeship. These four must work in a coordinated fashion for cooperative apprenticeship articulation to be successful. Each of the four systems will reap gains from cooperation; therefore, each should recognize the quid pro quo. Figure 2.1 presents this model.

ROLES AND RESPONSIBILITIES

The following provides an overview of the roles and responsibilities of the key representatives of each of these four systems, which are critical to cooperative apprenticeship. Secondary and postsecondary educators and business people must understand their mutual roles and responsibilities—and how these are to be carried out.

THE EMPLOYER SYSTEM

American business is faced with a number of economic, political, and sociological realities: (1) lack of available human resources trained to

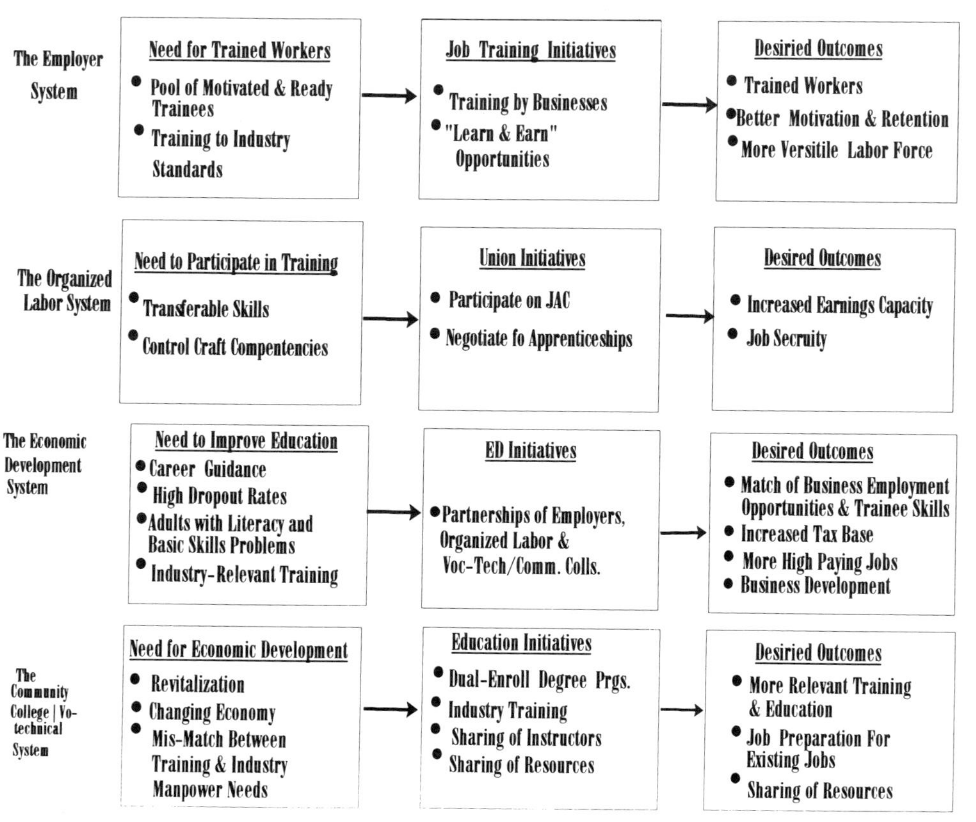

Figure 2.1 Four parallel systems.

business and industry needs, when and where such needs exist; (2) increasing numbers of retirees creating voids in skilled manpower in many businesses; (3) a perception of inadequate training and educational systems to support human resource development needs; and (4) increasing numbers of business failures due to rising costs to do business. As a result, it is fair to say that, certainly, out front in the pursuit of skilled workers is the employer—though most employers are not often willing to invest large sums of capital in training. In order to keep costs down and to maintain control over training, it is the employer who traditionally initiates apprenticeship; however, the vocational-technical school and community college education system can also serve as a catalyst to bring about such structured work-based training.

It is important to understand those barriers to apprenticeship that have been expressed by employers. Included is a fear of training workers in-house and then losing them to a competitor. A significant number of employers do sponsor formal training programs that could qualify as apprenticeships, but they refuse to register the programs with the government for various reasons, including the fear of interference in matters of business, as well as burdensome paperwork. However, all employers (not just employers working with organized labor) can benefit from sponsoring registered apprenticeship cooperatively with public educational institutions. For instance, some small-scale employers band together and establish mutually-funded programs, therefore spreading the cost of training over the entire partnership.

TEXTBOX 2.1

Preapprenticeship in the Maritime Industry

A very well recognized preapprenticeship program, which provides entre into Norfolk Shipbuilding's program, is conducted by the Tidewater Maritime Trades Foundation, called the Tidewater Maritime Training Institute (TMTI). TMTI has been in operation since 1982 and has successfully provided entry-level training for almost two dozen Tidewater area shipbuilding firms—entry-level to apprenticeship training. The Institute is unique in that it was founded by one local industry to meet its own needs, while serving to reduce unemployment, support the local economy, and upgrade the trainees' education and training.

TMTI is a nonprofit educational foundation (IRS Code 501c3) whose members are the area shipyards. TMTI trainees are taken into the Norfolk Shipbuilding program with credit for eighteen months of the three-year program.

[Adapted from Cantor, J. A. (1986, June). "The Job Training Act and Shipbuilding: A Model Partnership," *Personnel Journal,* pp. 118–125.]

THE ORGANIZED LABOR SYSTEM

Labor unions have a long-standing interest in apprenticeship and, in fact, have kept this form of training alive over a 200-year history. There is a consistency of purpose, however, between labor's interests and that of employers advocating the use of apprenticeship to provide broad and transferable training, thus maximizing a worker's service and earnings power over a career span. Cooperative apprenticeship satisfies some very important union objectives, the most important being the ability to control craft competence and productivity. From a collective bargaining perspective, uses of corporate investments in job skills development is a form of job security. By controlling the kinds of training, degree of specificity, and numbers of trainees, unions maximize their ability to attract and maintain contracts and keep their members versatile and adaptable to work changes. It also serves to prevent substitution of low-wage workers for journeyman workers.

Organized labor's interest in dual-enrollment processes for apprentice training in cooperation with community colleges dates back to the mid-1970s, with programs developed nationally by IBEW and IUOE. Community colleges have become training grounds for tomorrow's union leaders. Organized labor and professional industry groups together are beginning to recognize that the Associate degree is a basic necessity in many high-technology fields. The union is also recognizing that higher education will aid the worker in being a better and a more productive union member.

TEXTBOX 2.2

Dual Enrollment of Apprentices in College Degree Programs

Union apprenticeship programs have developed close ties to community colleges. In fact, the International Brotherhood of Electrical Workers national leadership supports dual enrollment of their apprentices. IBEW's Local #3 Empire State College program is a model for such articulation. Other programs for dual enrollment of apprentices in a community college degree program exist nationwide. Catonsville Community College in Maryland operates apprentice training. Catonsville has a full-time college apprentice program coordinator to interface with all unions in program delivery. Catonsville services IBEW, IUOE, and other union programs in an off-site facility dedicated to these kinds of programs. Other colleges such as Fox Valley Technical College support the development of apprenticeship instructors through instructor training at universities such as the University of Wisconsin at Stout.

THE ECONOMIC SYSTEM

A third and often unpredictable system at work, which interacts with business, organized labor, and the vocational-technical school and community college systems, is the economic system. All too often, jobs fail to exist for which the school or community college prepares trainees to work upon graduation, or such jobs disappear due to lack of business while trainees are in formal schooling. Unpredictable and uncontrollable economies make manpower development and training an imprecise science at best. The U.S. Bureau of Labor Statistics supports the fact that more than half of all new jobs created between 1984 and 2000 will require some education beyond high school. Almost one-third will be filled by college graduates. In fact, the large majority of employers consulted contend that the complex and changing marketplace will continually heighten the need for a more highly qualified entry-level labor force. Future workers will need to work in more complex work environments and demonstrate more complex and intricate tasks. A particular job title (e.g., automotive mechanic) may remain the same; however, it will change dramatically in content and skill requirements.

THE VOCATIONAL-TECHNICAL SCHOOL–COMMUNITY COLLEGE EDUCATION SYSTEM

Public education has certainly come under fire in recent years. Criticism has included an inability on the part of high schools to adequately educate youth for work and life. In the areas of job training, criticism includes inadequate job preparation skills, lack of literacy in general, and an inability for educators to work with or relate to local business and industry.

Much of this criticism is also levied at community colleges. While working more closely with local industry, some programs do tend to lose contact with the state of the art and become outdated, losing the ability to prepare competent workers. As we approach the year 2000, 75 percent of all job classifications will require two years of postsecondary education—best delivered by the community college. Labor, business, and civic leaders continue to seek out the community college to design curricula to link labor's needs and the world of work.

Many vocational-technical school systems sponsor preapprenticeship programs. In Connecticut, the apprenticeship law recognizes the value of preapprenticeship. The Connecticut Voc-Techs administer the preapprenticeship program in cooperation with the Connecticut Department of Labor. Likewise, in Georgia the vocational-technical school system works closely with the Georgia Department of Labor to administer apprenticeship programs.

Community colleges are unique institutions of higher education. They have the ability to design customized job training programs to the specifications of a firm or business; thus, they are capable of meeting varying needs of a community's business and industry and, simultaneously, of providing economic development support to their communities. Community colleges have played a role in supporting employer job training and apprenticeships for some time. Many, if not all, community colleges provide related trades training for employers. Community colleges have also been active in the apprentice training field for some time. While generally providing related classroom training, some have also designed programs to meet the increasing technological needs of the firms and industries they serve, as well as providing apprentice workers the higher education backgrounds (and, in some cases, basic skills education) that these workers will need. Many offer specific Associate degrees for an apprentice. Some award credit for hands-on apprenticeship work. Some tailor related theory—and even liberal arts and sciences courses—to the trade or occupational program.

Cooperative apprenticeship offers a solution by permitting the employer to place the trainee at an existing job and train that person "on-the-job" to requisite specifications. Organized labor also benefits from such a relationship. From an economic development perspective, a gainfully employed trainee filling a current job requirement and, at the same time, gaining education for tomorrow's career needs is an optimum situation. These interorganizational arrangements of employer, job training, organized labor, education, and economic development have the potential of bringing together these four traditionally conflicting systems: the employer system, organized labor system, vocational-technical school/community college system, and the economic development system.

TEXTBOX 2.3

Business and Education Cooperate for Apprenticeship

Strengthening the economic base of the U.S. maritime industry is a major concern to the shipbuilding community and our nation as a whole. Training is an integral part of development of the economic base of this industry. The Newport News Shipbuilding and Drydock Company operates a prestigious four-year apprenticeship program in twenty trade areas. Thomas Nelson Community College (TNCC) provides on-site related trade theory classes, advanced training, and general education courses to complement the program. All costs, including college-related course costs, are borne by the firm for the apprenticeship training; additional college general education costs are reimbursed by the firm upon successful completion of the college work towards an Associate degree. College work follows the apprenticeship training, after hours at the firm. Of noteworthiness, the apprenticeship itself is creditable towards a degree. Apprentice program completers will receive a TNCC Engineering Technology Certificate; upon completion of the additional college courses, the student can earn an Associate in Science degree in a technology area. About 20 percent of the apprentices complete the two-year college degree.

The Bath Iron Works program is somewhat different in method of delivery. Its close liaison with the Maine Maritime Academy (MMA) provides an opportunity for a Bath apprentice in any of five different marine design areas to earn an Associate degree after completion of 8,000 hours of instruction and as part of the apprenticeship work with the company. Trade and related education instructors are Maine Maritime Academy faculty, who work on-site at Bath Iron Works. The MMA and Bath Iron Works developed the program together and jointly operate it. Program options include electrical, HVAC, hull, piping, and structural specializations. About fifteen credits of the college curriculum are awarded for the technical hands-on portion of the apprenticeship.

Lastly, NASSCO's four-year apprenticeship program is supported by San Diego Community College. NASSCO's apprentices attend college-related instruction courses together with apprentices from a number of other firms in the area. Courses are taught on NASSCO's site or other area firm sites. They earn three units of college credit per semester, over eight semesters—attending classes two nights a week after work hours; however, to complete an Associate degree, they must complete additional arts and sciences courses on their own, after the apprenticeship period is completed.

THE GOVERNMENT'S ROLE

As already stated, the government oversees apprenticeship in a number of ways. In SAC states, the state department of labor assists firms to locate, recruit, and screen candidates for apprenticeships and then registers them. The U.S. Department of Labor's regional office does the same in non-SAC states. Both agencies then monitor apprentices while on the job and ensure that training is taking place to uniform occupational standards.

STUDENT NEEDS

For noncollege-bound youth or for those not planning an immediate college experience and particularly for those most vulnerable to dropping out before completing high school, formal apprenticeship training programs offer opportunities to bridge the gap between school and work. There is a growing belief among educational researchers that current school policies are poorly designed for integrating noncollege-bound youth into meaningful careers. Typically, schools and colleges have weak linkages with employers. Many of these youth will venture into jobs that will, at some point in time, require postsecondary education and training.

COMMUNICATING FOR COOPERATIVE APPRENTICESHIP LINKAGES

Apprenticeship requires securing placements for students in business and industry. Success in cooperative apprenticeship rests with good communications. This means developing an appreciation for how each party (vocational-technical school, community college, and business) operates. This chapter is about communication, the single, most important ingredient for developing and operating successful cooperative apprenticeship learning activities. Therefore, it is essential to understand how to establish successful working relationships through effective communication. Clear communication promotes cooperation between vocational-technical school and/or community college faculty and administrative staff, participating employers, and other government and community members.

EFFECTIVE COMMUNICATIONS PRINCIPLES

Effective cooperative apprenticeship requires sound communication. This results from an understanding of articulation. The following are eight principles that I have advocated (Cantor, 1994) as initially proposed by Dornsife (1992), for articulation:

- Establish a quid pro quo (mutual benefits to all parties must be obvious).
- Focus on mutually agreed-upon goals.
- Encourage faculty involvement.
- Foster leadership and commitment through clearly stated objectives.
- Ensure open, clear, and frequent dialogue.
- Develop performance-based curricula.
- Develop written cooperative apprenticeship agreements.
- Build relationships based on respect and trust.

Each of these will be discussed in detail.

MUTUAL BENEFITS TO ALL PARTIES MUST BE OBVIOUS

Effective communication in a cooperative apprenticeship arrangement happens when a quid pro quo exists, thus ensuring that all parties recognize those benefits to be derived by their participation. Those firms engaged or potentially engaged in cooperative apprenticeships, as well as the vocational-technical school and community college, must clearly see the benefits to be mutually derived by participation. This might be access to a potential workforce and a controlling hand in the training program design and delivery by the firm and/or access to modern up-to-date labs and experienced mentors by the school and/or college. Experience has demonstrated, though, that when all parties see the benefits potentially to be derived by their particular organizations, a willingness to commit the time and resources necessary for meaningful dialogue will begin to happen. How can this be done?

First and foremost, educators must understand the perspective of the participating firms. Analysis of successful cooperative apprenticeship partnerships indicates that those educators who can appreciate what is often termed a "private sector perspective" more often tend to conduct successful cooperative apprenticeship programs. These educators see the roles of schools and colleges as a community partnership team service provider with responsibility for carrying out the instructional goals adopted by the partnership entity. What is most important, however, is that educators understand that they cannot assume total program control. Hence, they are ready to deal with issues of accountability on the part of the educational institution and to negotiate and seek consensus when problems do arise. One important aspect of this role is a commitment to respond quickly when entering into an apprenticeship agreement, bringing the requisite classroom education and training on-line when and where required. These successful program participants make decisions quickly and keep their calendars very flexible.

TEXTBOX 2.4

Apprenticeship Communications and Advisory Committees

Apprenticeship program advisory committees are excellent mechanisms for communication. These committees exist at both the national and local levels. In many communities, they are hosted or arranged by the state department of labor, bureau of apprenticeship and training. The Springfield Fire Department's apprenticeship coordinator represents the fire department's apprenticeship program on one such committee and brings back ideas on program operations. At the national level, the National Apprenticeship Program for Firefighters has standing committees for coordination of standards and training, represented by other fire service organizations such as the Joint Council of Fire Service Organizations, the National Professional Qualifications Board, and the firefighter and fire officer unions.

Automotive manufacturers also have representatives who participate on school and college advisory committees. As a stipulation to participate and sponsor automotive program apprenticeships, most manufacturers insist that their organization's representative participate on the school's program advisory committee to ensure good solid communications. Additionally, in the automotive training milieu, instructor organizations such as the National Association of College Automotive Teachers meet regularly to exchange information and ideas.

In the maritime industry, a regular group of apprenticeship program operators meets nationally to discuss the state of the art in apprenticeship training. This committee is comprised of federal agency, military, commercial, and some international ship construction executives.

The construction industry is also represented on advisory committees. Each of the large national labor unions (e.g., IBEW and IUOE) has national joint apprenticeship committees (JACs) that oversee apprenticeship program standards and operations and that meet with state and local affiliates. In addition, these groups have representatives on school and college program committees where their apprentices are trained.

FOCUS ON MUTUALLY AGREED-UPON GOALS

Successful cooperative apprenticeship partners make the apprentice the top priority at all times. They made a decision to engage in a cooperative training venture based upon goals that they believe in. Hence, now they must focus on these program goals and not on issues of turf. Successful partners are willing to work for the common good, instead of to their own advantage. Success begins when these partners start with issues that all participating members can agree upon, and build on these. If and when problems occur (and they will), the partners will concentrate on the common goals and the good of the apprentice.

DEVELOP MODEST INITIAL GOALS

It is always best to start small and build to multiple numbers of cooperative apprenticeship placements. Often, the easiest place to start is with a single placement site, which you might already have, or develop it around an existing practicum or work experience activity. Concentrate administrative efforts and resources towards only one or two placements at a time and do the job well. The success and benefits from that first established apprenticeship placement will convince others and other programs to participate.

Employers of all sizes and types should be included in planning these partnerships. Large employers are usually better able to handle the added supervisory responsibility associated with vocational-technical school/college-to-work programs and generally have more on-the-job training slots to offer. Small businesses, on the other hand, often look to partnerships as a source of part-time workers. While learners might not receive as much structured training within small-employer contexts, they generally gain more varied work experiences. Smaller businesses also gain from partnership involvements by extending their influence and networks. They speak favorably about the apprenticeship program and share credit for its success. Think about forming an educational foundation to serve as a catalyst for interorganizational communications.

PROMOTE FACULTY INVOLVEMENT

In order to insure a successful cooperative apprenticeship program, as many faculty and staff members as possible (representing all programs that will have apprenticeship components) should be invited to participate in the meetings held for program planning and development. Wherever possible, members from business and government who will be instrumental in apprenticeship program operation should be invited. This will help all parties to get to know one another, understand each other, and respect one another. Meetings should rotate between any and all facilities so that instructors can better understand the other program. I have found that, in some industry programs, this happens very nicely. For instance, in the automotive industry, the car manufacturers use Voc-Tech and community college faculty to help develop the national curriculum materials. In the fire service, faculty work on national standards development committees.

STRIVE FOR ACCOUNTABILITY

Someone must be assigned the responsibility for keeping open channels of communication for all of the cooperative apprenticeship program partners. Vocational-technical school and/or community college administrators should jointly fund a single coordinator for all of the apprenticeship placements operating with the institution(s). Possibly, an industry person can recommend an appropriate person to fill this bill. All staff must have clearly defined responsibilities, and those who are coordinating the effort must clearly understand these responsibilities. For instance, if a particular partnership might involve sixteen apprenticeship placements, only one person should administer and operate it.

CLEARLY STATED OBJECTIVES FOSTER LEADERSHIP AND COMMITMENT

The key element for effective cooperative apprenticeship program development success is communication from the top of the organization. Secondary school administrators, as well as college deans and directors, must take part of the leadership responsibility for communicating institutional objectives and support for work-based learning through cooperative apprenticeship and should commit institutional resources towards that effort. Program coordinators and department chairs and faculty must also provide leadership to cooperative apprenticeship programs.

ENSURING EMPLOYER COMMITMENT

On the business side in successful and ongoing cooperative apprenticeship programs, employer commitment typically comes from the CEO or from, at least, a senior vice president within that firm or organization. Likewise, top elected officials on the public sector side and superintendents of vocational-technical schools and/or presidents of community colleges must also display a "front office" presence. A genuine commitment to making the work-based learning experience more meaningful, more relevant, and more effective for all learners must originate from the top of each organization if partnerships for cooperative apprenticeship are to be successful. These educational institution policymakers and leaders should make sure that the objectives and values of cooperative apprenticeship are understood and supported. This top-level commitment brings high visibility to the partnership.

While decision-making authority rests at upper levels, the responsibility for maintaining contacts, generating support, and making operational decisions must extend to the program operators—the program faculty and participating employers—for these programs to grow and develop. In this way, ownership of the apprenticeship program and pride in the outcomes are experienced throughout, and commitment is solidified.

COMMUNICATE THE OBJECTIVES

Before anything happens, again, there must be a comprehensive understanding among all regarding desired objectives and program outcomes. When developing these objectives, it is important that the short-term and long-term "customer" needs of each participating partner also be served. These expectations should be written and formalized through employer/ school agreements. Samples of these will be given in the following chapters.

Cooperative apprenticeship program partnerships should be guided by these basic principles:

- *Create motivation:* Cooperative apprenticeship job opportunities and work activities should motivate learners to complete both high school and college degree programs and become productive workers and citizens.

- *Ensure high standards:* Cooperative apprenticeship standards should be designed to allow learners to attain high workplace and academic goals.
- *Link work and learning:* Cooperative apprenticeship agreements should link Voc-Tech school and/or college classroom curriculum to work-site experience and learning.
- *Focus on employment and careers:* Cooperative apprenticeship placements should enhance the learners' prospects for immediate employment after leaving Voc-Tech school and/or college and/or for entry on a path that provides significant opportunity for continued education and career development.

ENSURE OPEN, CLEAR, AND FREQUENT DIALOGUE

The most successful cooperative apprenticeship partnerships are characterized by open and candid communication at all levels, often on a daily basis and regarding all aspects of the training program activity and its policy. New ideas are encouraged. If an idea is rejected, partners are urged (and expected) to rework it or describe the problems with it and offer suggestions about how it could be made useful.

Cooperative apprenticeship communication needs to be open, clear, and frequent. Communication needs to take place at all different levels and among counterparts in the participating firm or business and internally among all members of a vocational-technical school and/or community college. If all necessary participants do not know about and support a cooperative apprenticeship program development effort, it can't be successful. Industry education groups such as the National Association of College Automotive Teachers or the National Fire Service Instructors Association meet regularly—and this provides for open lines of communication.

This kind of behavior produces a sense of ownership and is a catalyst for contribution and commitment within all collaborating partner organizations. It fosters the desire to speak freely about the apprenticeship program and share credit for its success.

Furthermore, participating partners must understand that significant time is required for the formulation of stable, lasting cooperative apprenticeship partnerships. Creating strong, lasting programs requires a great deal of time. Most of this is *expensive* time because of the significant involvement and commitment required of a firm's and vocational-technical school and college's senior managers, especially during the initial implementation phases. Moreover, continuous redesign will be required throughout the life of the cooperative apprenticeship program in order to keep it operating efficiently. A good school/college-to-work partnership may take a decade to develop, with problems along the way.

INITIATE PERFORMANCE-BASED CURRICULA

Programs are easier to understand and to articulate when coursework and on-the-job learning activities are built around the performance objectives to be learned and mastered. With a performance-based apprenticeship, instructors and on-the-job supervisors and mentors can better coordinate the educational experiences for learners. A performance-based apprenticeship provides instructors the opportunity to have a common approach and a common language for planning. Develop *measurable learning objectives* for cooperative apprenticeship programs and specify these on apprenticeship agreements. Use existing industry standards, where such exist, and national standards project data, or enlist education's assistance to identify and develop standards for your program. More is said about this in Chapters 5 and 6.

DEVELOP WRITTEN COOPERATIVE APPRENTICESHIP AGREEMENTS

Register your program and your apprentices with a state or the Federal department of labor. Put all agreements in writing. Formally written agreements should be signed by the chief executive officer of the schools and/or colleges with the employers, and usually registered with the department of labor. Chapters 5 and 6 will also display and discuss these agreements.

BUILD RELATIONSHIPS ON RESPECT AND TRUST

Solid partnership relationships are built on respect and trust. Respect begins with involvement. This usually takes time and is developed through the process of working together on common goals.

Training partnerships should foster climates of negotiation and cooperation. Research indicates that, frequently, partnerships create independent oversight entities (such as joint apprenticeship program advisory committees) or seek the assistance of outside organizations (e.g., joint apprenticeship councils or economic development/labor groups) to function as brokers (Cantor, 1990). These third-party players serve as a catalyst to promote the program and reduce the impression that any one partner is serving a vested interest. Generally, the broker's role is to see that focus is held on learner needs and expected outcomes. By continually emphasizing learner needs and outcomes and by maintaining open dialogue, all partners inevitably look beyond their own self-interests. In some localities, nonprofit foundations comprised of members of all participating organizations foster this initial trust. Chapter 5 will deal with this further.

NOW, BEGIN THE PROCESS

The question of "who's in charge?" doesn't matter, according to those who have been successful in developing cooperative apprenticeships. Sometimes the lead is taken by the vocational-technical school or the community college; in other cases, it has been the employers who took the first step in exploring possible joint efforts. Either way, someone has to think about these questions:

- What businesses, organizations, and learners will benefit from these discussions?
- Who are the key players in each business?

STEP #1: SCHEDULE A PLANNING MEETING

The agenda for a planning meeting can be relatively simple but should include the following:

- Identify firms who might participate in cooperative apprenticeship.
- Establish cooperative apprenticeship program goals.
- Develop a time line and listing of tasks to meet goals.
- Set annual cooperative apprenticeship program objectives.

STEP #2: SECURE THE ENDORSEMENT OF CHIEF EXECUTIVE OFFICERS

Again, successful cooperative apprenticeship program development ultimately requires the policy-level approval of the chief executives at each institution (Voc-Tech school/community college and business) and the endorsement of each board of directors or trustees. A policy statement inserted in the formal decision-making process for each organization will give credence to all remaining steps. Even Voc-Tech schools and colleges that have taken the more informal approach to cooperative apprenticeship program development may eventually choose to have a set of written policies to guide staff who follow them in the future.

STEP #3: DEVELOP COORDINATING PROCESSES

A program coordinator begins the process of building and developing cooperative apprenticeship partnerships. The following steps should be performed:

- Identify one person to serve as each single apprenticeship program's coordinator.

- Identify those programs that will be tackled first.
- Develop formats for a prototype agreement that will fit the local situation (see Chapter 6).
- Discuss and identify faculty to spend release time in program development, building desired competencies that learners should master in the respective apprenticeship activities.
- Establish the mechanisms for maintaining consistent communication.
- Document all meetings and decisions to maintain a formal record of progress.

STEP #4: ORIENT STAFF MEMBERS

It will be up to the "middle managers" in the institution to orient staff to the possibilities and potential of cooperative apprenticeship:

- Emphasize the policy-level commitment of the chief executives and the respective advisory committees.
- Describe the process of the proposed apprenticeships.
- Provide adequate and detailed instructions and models to follow.
- Identify a facilitator (or chair) for meetings.
- Clarify how faculty and staff initiative enhance the possibility of success.

STEP #5: ARRANGE WORK SESSIONS

Cooperating instructors (and employers) will need to work out details of apprenticeship program offerings, course and work-site sequences, competency lists, and standards of performance as previously described. Provide opportunities to visit each other's firms and campuses, meet jointly and review goals and objectives, and discuss instructional strategies and resources.

STEP #6: COMPLETE DRAFT AGREEMENTS

The committee and program coordinator will need to decide if the draft agreements will need to be presented to the board of trustees for approval or if administrative approval will be sufficient.

STEP #7: IMPLEMENT THE PROCESS

All participants who work to implement a cooperative apprenticeship program should continually gather and share important information.

- Each faculty member must become an enthusiastic advocate for the process that has emerged.
- Particular attention should be given to sharing the new program activities.
- Data on the learners who begin to take advantage of the system need to be accumulated for program refinement and reporting.

STEP #8: REVIEW PROCESS ANNUALLY

Chapter 9 describes the review and evaluation process.

SUMMARY

Effective communication is essential for cooperative apprenticeship relationships to work. Vocational-technical school and community college faculty and staff must be able to relate to business and industry in order for cooperation to take place and for cooperative apprenticeship to happen. This chapter has presented the fundamentals of effective communication for linkages of cooperative apprenticeship programs with business and industry and the community. Subsequent chapters will describe how to apply these principles to various kinds of cooperative apprenticeship activities. Once lines of communication have been established, memos of understanding or agreements drafted, and specific competencies identified for which cooperative apprenticeship is appropriate, the next model step involves selecting work sites and matching learners to work sites. Chapter 7 will describe this step in detail.

REFERENCES

Cantor, J. A. (1994). *Cooperative Education and Experiential Learning.* Toronto, Ontario, Canada: Wall & Emerson.

Cantor, J. A. (1990). Job training and economic development initiatives. *Educational Evaluation and Policy Analysis,* 12(2):121–138.

Cantor, J. A. (1986, June). "The Job Training Act and Shipbuilding: A Model Partnership," *Personnel Journal,* pp. 118–125.

Cantor, J. A. (1985). A local industry solves its training needs: A cooperative training venture that works. Presentation to the National Conference on Technical Education, American Technical Education Association, Charleston, SC.

Dornsife, C. (1992). *Beyond Articulation: The Development of Tech Prep Programs.* Berkeley, CA: National Center for Research in Vocational Education.

Kendall, J. C., Duley, J. S., Little, T. V., Permaul, J. S., and Rubin, S. (1986). *Strengthening Experiential Education within Your Institution.* Raleigh, NC: National Society for Internship and Experiential Education.

Montgomery County Consortium for Vocational/Technical Education. (1990). *Final Report: Strategic Plan for Collaborative and Articulated Vocational/Technical Programs.* Montgomery, PA: Author.

Thomas, H. B. (1983). *Linkage between Vocationally Trained Participants and Industry Registered Apprenticeship Programs: A Study of Barriers and Facilitators, Final Report,* Tallahassee, FL.

U.S. Department of Labor, Employment & Training Administration. (1992). *School-to-Work Connections: Formulas for Success.* Washington, D.C.: Author.

U.S. Department of Labor, Employment & Training Administration. (1989). *Work-Based Learning: Training America's Workers.* Washington, D.C.: Author.

U.S. Department of Labor, Employment & Training Administration. (1988). *Building a Quality Workforce: A Joint Initiative of the U.S. Departments of Labor, Education and Commerce.* Washington, D.C.: Author.

From Doing a Needs Assessment to Deciding When Apprenticeship Is Appropriate

THE purpose of this chapter is to provide insights into decision making about using apprenticeship as an instructional methodology for your students or trainees. However, if you are an instructor or educational administrator, you will not be able to offer an apprenticeship program without selling the idea to businesses in your community, as well as to students and other faculty within your institution. Therefore part of this chapter is about marketing apprenticeship to firms, businesses, and the community-at-large, as well as to your students, faculty, and administrators within your institution. Through a needs assessment, you will be able to gather information to help these people realize that apprenticeship can help to meet their needs and goals.

WHY NEEDS ASSESSMENT?

Training program development and delivery costs money, and in times of tight budgets, school administrators and business managers are wary of initiating new undertakings without appropriate justification. Therefore, a solid needs assessment is an imperative before undertaking development of apprenticeship training programs. A model for needs assessment is provided.

A THREE-STEP NEEDS ASSESSMENT PROCESS

Kaufman's (1976) three-step approach to needs assessment is very applicable to cooperative apprenticeship. Kaufman's three levels of assessment are MEGA, MACRO, and MICRO. For our purposes, the MEGA level is considered to be the business level. Here, we need to make determinations about where and how apprenticeship training can meet business and/or community needs for trained workers. The next level, the MACRO level, is where we look at the needs of our vocational-technical schools and community colleges and their various programs. We ask the question: How can apprenticeship help to meet instructional needs within our educational institutions and their programs? Finally, there is the MICRO level—the level concerned with our students' needs—and how apprenticeship can meet their immediate and long-term needs for training. We'll take a closer look at each of these levels of needs, beginning in reverse order with the MICRO level—the student (see Figure 3.1).

MICRO LEVEL NEEDS ASSESSMENT

Why do students (or young people) want to be apprentices? To begin assessing the value of apprenticeship, let's look at what apprenticeship has to offer from a young person's point of view.

- All too many young people do not have clear visions of what careers exist for them—or for that matter, what careers offer long-term stability and growth in the economy. Since apprenticeship is based upon industry needs, those apprenticeable occupations tend to be more stable and growth oriented.
- Apprentices generally enjoy a high level of job security and are assured of increasing wage levels through the term of their apprenticeship. Entering an apprenticeship is a long-term commitment for the individual as well. It is not only the commitment to a specific employer, but also the commitment to attend school both during the day and in the evenings, as spelled out in the indenture agreement. Students invest significantly and stand to gain a great deal through apprenticeship.
- Young people need to be motivated to learn. As a hands-on learning activity and one in which students "earn and learn," student motivation is high. This makes it especially attractive for young people engaged in formal secondary school indentured preapprenticeships. Learners who cannot afford to be away from gainful employment while preparing for a new profession or career can "learn and earn" through cooperative apprenticeship. Apprenticeship is an excellent example of experiential learning.
- Many young people need to experience success in learning, especially those who have experienced nothing but a continuing failure chain. Apprentices have an opportunity to learn skills, perfect them over time, interrelate theory and skills, and learn to function within the work and organizational environment. The firm is able to supervise the trainees' skills development. Therefore, for a youth seeking a viable and profitable career:

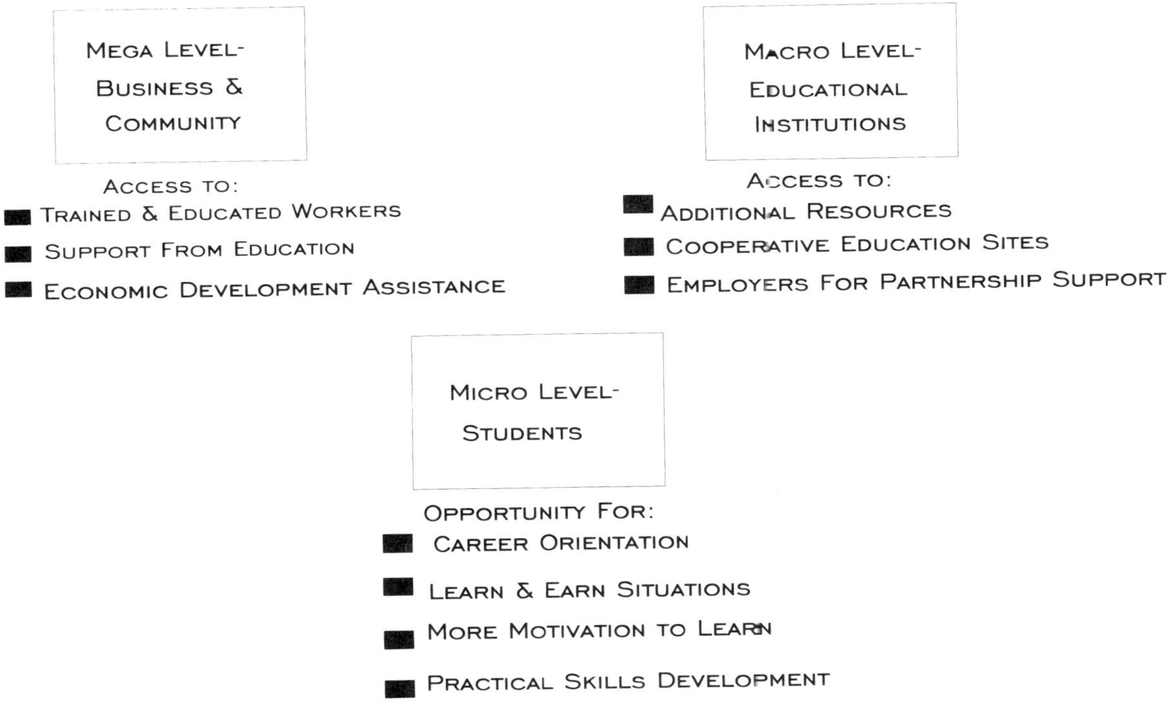

Mega Level-
Business &
Community

Access to:

- Trained & Educated Workers
- Support From Education
- Economic Development Assistance

Macro Level-
Educational
Institutions

Access to:

- Additional Resources
- Cooperative Education Sites
- Employers For Partnership Support

Micro Level-
Students

Opportunity For:

- Career Orientation
- Learn & Earn Situations
- More Motivation to Learn
- Practical Skills Development

Figure 3.1 Needs assessment: a three-phase process.

(1) Serving an apprenticeship provides a learner with a lifetime skill and a comprehensive knowledge of one's trade.
(2) Apprenticeship training enhances economic security for the individual, since graduate apprentices are often the first to be promoted and are among the last laid off.
(3) Graduate apprentices are well prepared for advancement to supervisory positions because they have proven themselves throughout their apprenticeship.
(4) The skills apprentices learn are transferable from one employer to another and generally from one area of the country to another.

Making a Decision

In assessing apprenticeship for specific students, if a fair number of the above cited benefits appear to match the student's goals and desires, then an apprenticeship placement in a specific occupational area of interest to that student would be appropriate.

MACRO LEVEL NEEDS ASSESSMENT

What does apprenticeship as an instructional tool offer to the vocational-technical school and/or community college? There are numerous benefits accrued to educators in vocational-technical schools and/or community colleges as a result of cooperating with employers to develop and conduct apprenticeship training. To determine if apprenticeship is an appropriate instructional method for you to use to deliver instruction in a particular occupational or technical area, ask yourself the following questions:

- Do you need to provide more motivating (and perhaps meaningful) learning experiences for your students?
- Can your educational program benefit from recruitment of more demographically diverse students?
- Do you, as well as other faculty in your educational institution, wish to ensure that your career and occupational programs are kept technically up-to-date?
- Do you and your colleagues wish to establish closer ties to your technical counterparts in business and industry (and civic and economic communities)?
- Can you use the assistance of local business and industry people to serve as curriculum resource people and mentors to your students?
- Can your institution's various educational programs benefit from

additional financial assistance often realized through apprenticeship program participation?
- Can the institutions and educational programs benefit from additional kinds of resources and benefits such as access to modern up-to-date laboratories and equipment?
- Can your educational institution and program stand to benefit from easier placements of students into employment after program completion?

Making a Decision

When is apprenticeship appropriate? You should now address the above questions. Dialogue with your students and their parents. Community economic development and human resource development agencies, chambers of commerce, and other organizations can help you to determine to what extent needs exist for workplace training opportunities in futuristic careers and occupations. Hopefully, as part of your regular and ongoing program needs assessments, you are already doing much of this.

If the answers to most of these questions are yes, then registered cooperative apprenticeship might be a viable instructional option for your program(s). From this basis, you can now meet with prospective employers to discuss forming cooperative relationships with them for apprenticeship training.

And finally . . .

MEGA LEVEL NEEDS ASSESSMENT

What does apprenticeship offer to local business and industry and the community at large? Discussions you will have with these prospective employers will help answer the following questions:

- Do employers view the current pool of entry-level workers as insufficiently prepared in basic work-related skills?
- Are employers concerned that job competencies of entry-level workers are deficient?
- Are employers indicating that basic reading and mathematics skills, all too often lacking in young people, are essential to job success?
- Are employers also indicating that entry workers need basic workplace skills, including flexibility and adaptability, problem solving, self-direction and initiative, and attitudes and work habits?
- Can the firm benefit from training through apprenticeship to mitigate or reduce the effects of turnover?

Do graduate apprentices recognize the investment the employer has made in them and the investment they themselves have made in learning about the company and its products?

- Can the employer benefit from highly motivated workers who are willing to spend extra hours in classroom training in order to learn the theoretical bases of their trade?
- Can the employer benefit from more highly productive workers who know the reasons behind the steps they must take in their work processes?
- Can the employer benefit from technologically up-to-date workers?
- Do business and community leaders see a need to provide a bridge from school to work for young people?
- Do community and business leaders see a need to support education in its efforts to prepare competent workers?

And what about organized labor?

- Can the employees' union benefit from cooperatively sponsoring apprenticeship training? Will it help to attract career-minded employees to the trade or occupation?

Making a Decision

Many employers will see the numerous benefits of apprenticeship. These employers (and perhaps community-based organizations, as well) will become the sponsors of cooperative apprenticeship work sites. For instance, the automotive industry was one of the earliest industries to accept cooperative education and apprenticeship program development (see Textbox 3.1).

TEXTBOX 3.1

Toyota's Program Involvement

Toyota Motor Corp. offers the following as rationale for their program involvement:

(1) There is increasing difficulty finding technicians who can diagnose and successfully repair sophisticated "hi-tech" vehicles.

(2) Rapid technological changes in motor vehicles result in technicians working harder to stay current with technology.

(3) Learners enrolled in traditional community college technical education programs lacked exposure to late-model vehicles and current automotive technology. Their skills are inadequate for success in the highly technical new car dealership environment.

(4) The high cost of state-of-the-art training aids and current levels of instruction are falling short of meeting industry's educational standards.

(5) Current methods of implementing cooperative work experience programs

were not designed to provide opportunities for learners to learn from experienced technicians.

(6) Training programs are not placing enough emphasis on diagnosis, which results in increased comebacks and customer dissatisfaction.

(7) There is a negative perception in many communities about the career of an automotive technician.

(8) With the growth in the number of dealerships and vehicle sales, Toyota needs to find qualified entry-level management candidates to fulfill this need.

[Adapted from Toyota Motor Corp., T-TEN Literature (1991).]

FINALIZING A NEEDS ASSESSMENT

At this point, you will bring the information and findings of each of the three levels of needs assessment together to finalize your decision making about the appropriateness of apprenticeship for your students. Determining apprenticeship as an appropriate method of training requires an understanding of all of the foregoing factors that contribute to making it successful as an instructional tool.

(1) You will have identified a real and documented need existing in a business or in the community for workers with specific skills and knowledge in a defined occupation.

(2) You will have identified an occupation or technical profession that is well enough defined for training in terms of performance standards.

(3) You will have identified resources available to provide structured work

TABLE 3.1. **Industry-Developed Cooperative Education/Apprenticeship Programs.**

Organization Program	Associate Degree
U.S. Office of Personnel Management	Office Administration
U.S. Navy	Shipyard Technician
National Machine Tool Association	Toolmaker/Machinist
National Automobile Dealers Association	Automotive Technician Automotive Management
Toyota Motor Corp.	Automotive Technician
International Brotherhood of Electrical Workers	Electrical Technician
International Union of Operating Engineers	Heavy Equipment Technician

experiences, including mentors or master employees to serve as on-the-job trainers.

(4) You will have identified students desiring a formal learn-and-earn situation.

(5) You will have identified learning objectives that indicate that instruction should occur in both the classroom and the work environment.

(6) You will have identified local firms needing training support from the vocational-technical school and/or community college in the form of recruitment of trainees, classrooms, curriculum and course development, and possibly training funds.

TEXTBOX 3.2

Corning Incorporated

A major company recognizes the continued benefits of apprenticeship training. This New York–based firm has been using apprenticeship to train its workers in over a dozen different trades, including some high-tech areas. It recognized that a partnership with higher education could bring the old system into the new era. Through an arrangement with Corning Community College, the firm is now providing apprenticeship training in fields such as instrument/electrician, machinist, and electrical trades, which require higher order cognitive skills. College credit towards an Associate degree is awarded by the college.

[Adapted from *Technical & Skills Training*, August/September (1993), p. 34.]

In addition, a correct match of the work sites for work-based training and cooperating employers must also exist, and of course, the right students must be identified for apprenticeship training. Again, these will be discussed later in Chapter 8.

If you have determined the foregoing to exist, you will then proceed to develop a cooperative apprenticeship. Table 3.1, on p. 49, lists some of the prominent firms and organizations that have developed cooperative education and apprenticeship programs to address their need for trained workers.

FINALLY MARKETING COOPERATIVE APPRENTICESHIP

If you are sold on cooperative apprenticeship as a viable instructional tool to supplement and complement your school or college's career and technical or occupational programs, then here are some of the arguments to make when meeting with and answering questions of the businesspeople and other community-based organizations where work sites and training opportunities might be available and appropriate.

How will apprenticeship training make my employees better workers? Today's, as well as tomorrow's employees must be more creative, able to work in teams, solve problems, and make decisions on their own. They must also be able to continually educate and upgrade themselves as technology advances. In order to keep up, education must be responsive to rapidly changing workforce and workplace needs—through business and industry articulation and linkages such as cooperative apprenticeship. During an apprenticeship, the hands-on, practical aspects of the job are mastered as apprentices are rotated through all phases of the particular occupation according to a written agreement stipulating the period of apprenticeship and the objectives to be mastered (performance expectations). The related instruction continues throughout the apprenticeship term and provides the opportunity to consider in depth the underlying principles of the job

As an employer, what will this cost me? The employer will need to know that apprentices are compensated at predetermined rates of pay, depending on the state's minimum wage law. Apprentices will also be covered by applicable unemployment compensation insurance. They are bona fide employees who are employed to "work and learn" a trade or occupation. Since apprentices are full-time employees of the company in which they are apprenticed, the system includes a pay scale for them. Required duration of training ranges from one to six years, depending on the specific occupation.

Apprentices earn lower training wages until they are able to perform higher levels of skills. JACs offset wages paid by the firms. Usually, the wage scale averages about half the journeyman rate in a particular trade or occupation. It increases progressively as the apprentices successfully master the training objectives and program segments. Herein is one very significant benefit of apprenticeship as a means of cooperative education.

As an employer, what will I have to do? Employers must realize that they will be making a long-term commitment to an apprentice. They will need to devote time and money (in the form of a wage) to the apprentice over what is usually a three- to four-year term of employment and training. These typical combined arrangements are covered in the standard apprenticeship indenturing agreement between learners (apprentice) and employers—as well as state departments of labor, if an apprentice is formally registered. These arrangements ensure employability for the learner and guarantee competent workers for the firm or industry.

They must realize that this training is to be based on a predetermined schedule of specific tasks to be taught and learned by the apprentice. They must be able to predict a long-term period of employment for the apprentice, and in these trying economic times, it is difficult for most businesses, regardless of size, to make long-term predictions of personnel need. However, you, as the educator, must sell the employer on investments in human resource potential for business growth—and this is what is meant in Chapter 2 as a "business perspective." When employers decide to sponsor an apprenticeship, they make a long-term commitment to an individual's training. They take this step to improve their company's profitability and prospects for growth, and employers have found that they gain from sponsoring apprenticeships in a variety of ways. Employers also benefit from apprenticeship by having a highly skilled workforce that can adapt to the shifting demands of new technologies.

TEXTBOX 3.3

Apprenticeship Reborn

Downsizing of major industry causes many areas of the firm to be discarded—often those seen as frills. When International Harvester reorganized as Navistar International Transportation Corporation and did its downsizing in the late 1970s, the apprenticeship training program was one such casualty. However, soon after, the firm saw the need to restart the program to capture the disappearing trade skills of its "graying talents" who were approaching retirement.

The firm negotiated a new four-year apprenticeship system with the United Auto Workers. Clark State Community College provides the related classroom instruction.

[Adapted from Pooler, Doug. (1993). In *Technical & Skills Training,* August/September, pp. 29–30.]

Will I be able to select and control the apprentice? The employer is able to select the apprentice to be hired, just as one would be able to select any other worker. Herein a quid pro quo can exist between you as the "education system" and employers. Educational institutions can help employers by providing a central organization for recruitment of potential trainable talent from which employers can join in the interview and screening processes for apprentices. Make this offer, and you are well on the way to a good relationship complete with open lines of communication, as was discussed in Chapter 2.

Secondly, as an employer, they have a right to control the apprentice and, if the relationship does not work out, terminate the employment relationship. However, assure the employer that you will be helping to counsel and evaluate the apprentice and, hopefully, to head off any problems before they get to that troublesome proportion.

What occupations are appropriate for apprenticeship? In answer to the question of what occupations or jobs are appropriate for apprenticeship, some state departments of labor will honestly say that just about any occupation that can be technically defined is apprenticeable. In SAC states, the state apprenticeship council will sit in final judgement on an employer's request for registration of an apprentice in a "nontraditional" occupation. However, the U.S. Department of Labor lists approximately 700 occupations—trades or crafts—for which an apprenticeship training system exists.

APPRENTICESHIP OCCUPATIONS

What makes one occupation suitable for apprenticeship and another more appropriate for another kind of training? There are several criteria to determine if an occupation should be recognized as apprenticeable. Generally, the occupation must:

- involve technical (though not necessarily manual or mechanical) skills
- be customarily learned in a practical way through on-the-job training and classroom instruction
- be clearly identified and recognized throughout industry

Since 1911, over 300 occupations have been recognized as apprenticeable. New trades are constantly added as the need arises. Apprenticeship training programs have been developed for many jobs that were once self-taught through trial and error or informally taught through on-the-job training alone (see Appendix B).

The apprenticeship approach has been especially successful in newer occupations in skilled maintenance and in the expanding service industries. In the emerging environmental control and energy fields, new workers are being trained to treat water, air, and wastes, to install and service environmental control systems, and to work in a variety of other technical jobs. Chapters 5 and 6 on program development will describe how to initiate a new apprenticeable area.

This does not prevent any particular industry or organization from adding to that list. Typically, occupations such as firefighter, automotive technician, X-ray technician, printer, police officer, and water treatment operator are included on the list. However, as new requirements emerge in an industry, new titles are added. For instance, in the fire and emergency services area, emergency medical technicians are trained by apprenticeship.

In summary, the goals and objectives of cooperative apprenticeship programs are to do the following:

- Establish a program that benefits the entire community (students, employers, government).

- Develop postsecondary training capable of providing the necessary training to meet the current demands of business for entry-level workers.
- Enable college faculty to update their skills and learn new skills from business and industry.
- Develop courses that emphasize creative problem-solving and diagnostic skills.
- Develop programs that enable learners to work side by side with experienced master craftsmen and technicians.
- Provide a learning environment that stresses professionalism and integrity.
- Establish a path for learners to continue their education at the college/university level, leading to management positions.

REFERENCE

Kaufman, R. A. (1976). *Needs Assessment: What It Is and How to Do It.* San Diego, CA: University Consortium of Instructional Development and Technology.

Preapprenticeship and Youth Apprenticeship Programs: A School-to-Work Concept

THIS chapter describes how apprenticeship as structured workplace training can begin during a student's high school years of study. Furthermore, this chapter discusses how partnerships of vocational-technical schools and community colleges can be encouraged with business and industry, organized labor, and government. Through such partnerships, successful apprenticeship training programs can be developed and delivered.

A RATIONALE

Ideally, career education should begin early in a young person's life. At each and every training conference I have attended in recent years, the message heard from business representatives was the same—our nation's youth fail to recognize the value of career training and tend to lack direction in terms of their life's work.

Fundamental to career education is a solid foundation in basic communications and computational and reasoning skills. Too many of those youth aspiring to tomorrow's high-paid technical occupations are ill-prepared for training in those occupations and careers due to a poor foundation in basic skills, science, and math. Furthermore, too many of our young people are ill-prepared to continue in higher education past high school.

Our nation is grappling with the need for a more comprehensive youth training and education system for American business and industry. This system undoubtedly will include cooperative preapprenticeship. As outlined in the Federal Committee on Apprenticeship position paper (1992), our nation's youth need to see a clear connection between school and work. This mandates partnerships between education and business, such as preapprenticeship. Through exposure to these career opportunities and options, youth can better prepare for continuing education and training upon high school graduation. Employers must share in the responsibility of training and educating tomorrow's skilled workforce through a joining of work-based and school-based learning. Cooperative preapprenticeship can serve as a major school-to-work transition strategy in this system.

At this point, you have made a decision to use preapprenticeship for training in your secondary school program based upon the needs assessment in Chapter 3. Now the task is to develop a youth apprenticeship or preapprenticeship. Following are some steps to guide you.

DESIGNING AND DEVELOPING PREAPPRENTICESHIP PROGRAMS

BASIC COMPONENTS

Preapprenticeships are typically established for students who have reached junior- or senior-year status in the secondary school. The student, employer, and parent enter into a written agreement (in some states, a registration agreement) specifying the kind of work and working conditions to which the student will be exposed.

The student works on a part-time basis during the school year and, oftentimes, full-time during the summer. Wages, which are specified in the agreement, must conform to state wage and hour laws. The student apprentice's work is supervised by the employer. Successful completion of the preapprenticeship is acknowledged by a certificate of mastery.

The student apprentice's related academic and technical education is coordinated by the school to connect work-based and school-based learning. After graduating from high school, the student should then be able to move into adult apprenticeship and continue an education at the community college. In some states, a portion of the hours accumulated in the preapprenticeship is credited to the adult apprenticeship.

A MODEL

Preapprenticeship programs are generally multifaceted (see Figure 4.1). After all, at the secondary school level, you are attempting to accomplish several educational goals through work-based education and training. Project ProTech in Boston has developed an outstanding model preapprenticeship program worth reviewing. It is best described in three phases. Initially, the student is exposed to career exploration. It is a known fact that all too many youngsters have little career information at the high school level; therefore, the program needs to fill that void. Generally, in programs that have been deemed a success, the student's junior year, first semester, is devoted to career exploration—rotating through a number of experiences on-the-job and perhaps shadowing incumbent workers. These experiences

Figure 4.1 A model for preapprenticeship.

take approximately ten hours per week of the student's time; therefore, it is a serious venture. Additionally, the students attend core academic classes (with tutoring help if necessary) to give them a well-rounded educational experience. The high school curriculum must be tailored with specific classes to make the preapprenticeship meaningful for participating students. These classes must connect the schoolwork with the work-site rotations. During the second half of the junior year, students who have met grade and attendance requirements are provided with part-time employment in their chosen fields, with a specific mentor assigned to each student. The summer phase following the eleventh grade will consist of eight-week full-time jobs in the students' chosen areas.

The second phase is usually a specific skills training phase. Here, the student has made a conscious career decision. For the remaining two senior-year semesters, the student will be developing specific work skills based upon long-term employment and educational goals. In this phase, the students are clustered in core academic classes with enriched academics for those students following a college prep curriculum (going on to the community college). Tutoring continues as needed. Senior-level students serve as peer counselors for junior-year students.

There are three types of skills training that can happen. Type 1 assumes that two or three semesters is adequate to prepare a student for an entry-level position after high school graduation. These students hold a part-time job throughout the school year, with a progression of responsibilities. Type 2 is reserved for positions that require a six- to eighteen-month period of education and training post–high school (e.g., community college certificate program). And Type 3 requires two to four years of college training after high school (e.g., AS/BS level). For college-bound students, college counseling is provided, including college application and financial aid assistance. The community college should establish a process to permit seamless transition into the college program for the high school graduates. Programs of this type exist nationwide, including the MechTech Program with Community College of Rhode Island and Catonsville Community College (see Textboxes below).

The third phase of this program is termed employment and post–high school training. Here, mentoring continues and tutorial assistance is provided for year three. Case management would also be provided for one year for those who went on directly to the workforce.

ESTABLISH A PARTNERSHIP

Program design continues with the establishing of a partnership. The partnership should consist of all of the necessary partners for your situation.

THE MUTUAL BENEFITS

Educational partners have a unique opportunity to establish an early communications link between business and industry and their schools and colleges. Through cooperative preapprenticeship, educators can offer students opportunities to apply the knowledge acquired in the classroom and master basic skills in their chosen or prospective trades and occupations. These students will have an opportunity to learn and earn—hence, profit from useful part-time jobs.

For the employer, cooperative preapprenticeship fosters an opportunity to train students before they enter the permanent workforce, thus gaining a chance to test out workers and reduce turnover. This will reduce the cost of training. This is also a proven method to attract adequate numbers of qualified job applicants.

If your program is to be a joint effort of a number of schools, such as in the Western Connecticut Health Careers Partnership (see Textbox 4.1) or the ProTech Model (Textbox 4.2), then the partnership should consist of:

(1) Employers (e.g., health careers—hospitals, nursing homes, community healthcare agencies)
(2) Superintendents of all of the participating school districts
(3) Teachers of appropriate occupations
(4) Representatives of all of the Voc-Tech or high school guidance departments
(5) Community college and senior college representatives if the program is to have upward career mobility for the students
(6) A single program coordinator who will be the point of contact for the program (e.g., private industry council)

TEXTBOX 4.1

Western Connecticut Health Careers Apprenticeship

The Western Connecticut Superintendents' Medical Apprenticeship Project was developed to provide students with the opportunity to combine education and job training. Supported by the Workforce Development Fund (JTPA), area schools, hospitals, community healthcare providers and local colleges and universities are working to assist students in developing specific work skills that will increase their employability in the medical field and/or facilitate entry into a post–high school training and educational program.

Junior- and senior-year students from eleven area school systems are selected to participate in the program. Employers have the final say on placement of these students. Students are supported through mentors from the participating university, as well as project staff. Students are expected to conform to employer's work standards, and on-the-job supervision is shared among the employer, project staff, and mentors. Students will receive training in confidentiality, fire, safety, and infection control prior to starting placement.

Employers are asked to identify positions that have a strong likelihood of offering employment after training and that will lead to positions where entry-level starting salaries are better than minimum wage. Each student will spend two years in the program. The first semester will be spent exploring career options in the medical field. The student will spend this semester shadowing healthcare workers in the field and spend some time in hands-on activities. The last three semesters will be spent working at one site. Students work between eight and ten hours per week. This schedule is tailored to fit around each student's high school schedule.

(Adapted from in-person interviews and promotional materials.)

If the program is to be a single apprenticeship for a single student, such as in the state of Connecticut's registered preapprenticeship program (see Textbox 4.3), then the partners will be:

(1) The state department of labor apprenticeship representative
(2) The high school instructor
(3) The high school guidance counselor
(4) The employer

TEXTBOX 4.2

Project ProTech

ProTech is a youth apprenticeship program combining school and work-based learning. Its aim is to prepare youth for careers in healthcare, financial services, and utilities/communications industries. The program spans the junior and senior years of high school and at least two years of college for the youth apprentice.

Project ProTech/Healthcare is a Boston Private Industry Council (PIC) collaborative effort with area hospitals, Boston high schools, community colleges, and Boston area healthcare organizations. It was initially funded by the U.S. Department of Labor—JTPA. It was developed to provide students of high school age with job skills in an area of high need—selected healthcare technician fields. In its fourth year of operation, the program provides hands-on skills training with classroom education—bridging school to work.

Junior-year high school students are recruited into the program. Students also attend challenging specially designed courses in the sciences (biology and chemistry) and math. They must adjust their study habits to meet this challenge. In their senior year of study, some advanced courses are creditable for advanced placement at participating community colleges.

Participating hospitals provide rotations and on-the-job training for the students. Hospital staff act as mentors and counselors. The rotations expose kids to target occupations to help them understand what specific careers are like.

ProTech project counselors help students link the high school classroom and the work site. Through seminars, they begin to make career choices. Through interfacing with supervisors, work habits are stressed.

Students benefit by having real jobs part-time during the school year and full-time in summers. These real jobs lead to real careers.

The quid pro quo is that the hospitals are training their workforce of the future. The educational system is turning out a supply of enough good health career workers. However, the students' personal growth is probably the most beneficial outcome inasmuch as it helps the students figure out what they want to do with the rest of their lives. It has proven to be a confidence builder, to instill responsibility, to help build values, to set goals, and to develop a sense of self-worth. It has been found that parental involvement is essential to program success.

In addition to financial services, banking and insurance, ultimately, the PIC is looking towards ten industry clusters by the year 2000, in conjunction with the Boston Public Schools.

(Adapted from PIC-ProTech promotional materials.)

PREPLANNING

An initial step in the design of all such programs is to develop the operating guidelines and lines of communications. This is best done in a preplanning meeting, covering the communication items discussed in Chapter 2. For instance, in the Western Connecticut Health Careers Program, the initial meetings were at the superintendent's level, in which the superintendent who felt most strongly about preapprenticeship as a means to provide students with career direction and motivating experiential learning experiences outlined his vision to the other ten superintendents and discussed how to operationalize the program. Follow-up meetings were with that group and other health career employers recruited by the group for participation. Once the preliminary buy-in by the chief educational officers was accomplished, then a steering committee of key and principal players (per above) was convened. This group deals with the program components and operating procedures.

TEXTBOX 4.3

MechTech, Inc.

The conceptual basis for MechTech is what cooperative apprenticeship is all about. The impetus for this cooperative apprenticeship training venture was the plight of the small manufacturer in New England, who felt the impact of the shrinking labor pool, retirements of skilled workers, and diminished employee skills. To meet this demand for skilled labor, a consortium of manufacturers to unite industry and education was formed and incorporated as a nonprofit corporation called MechTech. The corporation is led by a board of directors, an administrator, and staff. They set student wages, benefits, company charges; establish written policies; promote the program; and recruit students and participating firms.

Its mission is rather simple: to conduct an apprenticeship program that is affordable to all industry involved and of high quality to the apprentice. The program is four years in length, 8,000 hours of prescribed training in the precision metalworking and moldmaking industries. Education encompasses mathematics and sciences, communications and the arts, and the technologies associated with the metalworking industry. In place of the traditional 144 hours of related theory instruction, the apprentice pursues an accredited Associate degree in related theory for the industry.

The program is almost ten years old. Participating firms are engaged in machining, toolmaking or moldmaking, and/or associated areas as quality control, engineering, or CAD/CAM. They have a willingness to train MechTech students and ensure that the student will remain with the program until completion.

MechTech, Inc. is the employer of record. Participating apprentices work for MechTech during the apprenticeship and are placed at cooperating work sites. A rotation concept is used, and the participating firms provide the practical work atmosphere for the on-the-job portion of the student's training for predetermined periods of time. Apprentices receive a predetermined and competitive wage, medical insurance, and industry standard paid holiday and vacation benefits. There is also a reimbursement program for college tuition expenses.

Participating employers provide a written appraisal of the student's work after each rotation.

PLAN PROGRAM COMPONENTS

The next step in a preapprenticeship program design is to hold a planning meeting of all key players to discuss program components. As discussed in Chapter 2, sometimes it is best to hold this meeting at a site neutral to any of the key partners to ensure that all realize that this effort is a group effort, not one owned by any one partner. Sometimes, this is not practical, and if this is the case, the meeting sites should vary from partner to partner. All parties should realize benefits from participation. The planning meeting should be conducted in the manner described in Chapter 2 and cover the following major points.

Establish the *program's goals and objectives.* Again these goals and objectives should emanate from your needs assessment. They will probably include:

(1) To provide the students with career guidance
(2) To provide occupational information in the fields that the program will cover

(3) To allow students to gain useful occupational skills suitable for entry-level positions in a chosen career
(4) To assist students in their academic skills mastery
(5) To allow students to learn while earning money
(6) To permit students to study in a field for which they intend to continue as a career or on to college (and perhaps earn college credit)

These goals and objectives should be written down and agreed upon by all partners.

Now, *develop specific roles and responsibilities* for each partner, which include the following.

For the Program Coordinator

The responsibility of overall program coordination should be agreed upon by all partners (per Chapter 2). This single person will represent the partnership and have the authority to make the decisions agreed upon by the group. This person might be an employee of one of the partner organizations or be in a shared arrangement or be an employee of a separate organization formed by the partner organizations, as in the case of many educational foundations formed to provide school-to-work education and training. There is a great advantage to this kind of organization as will be discussed later in the book (again, see Textbox 4.3).

The program coordinator will work with the school counselors and teachers to:

- recruit students
- recruit employers in each community
- orient students and their parents
- plan curricula to support the students' academic goals
- resolve problems arising on the job
- evaluate students

The coordinator will also be responsible for holding at least weekly seminars for all preapprentices to discuss work-based topics such as employee behavior, responsibility, career information, and common work-related and specific problems and solutions to the problems.

For the Employers

There should be very specific kinds of roles and responsibilities for employers, which are established and agreed upon, in order for the work-site training to be meaningful to the student-apprentice. These responsibilities should be reduced to writing with job descriptions and indenturing agreements developed by the partnership. Responsibilities will include the following:

- Participate in the recruitment and selection of students.
- Interview and select potential high school preapprentices for the firm.
- Provide an appropriate and safe work environment for the apprentice.
- Provide a work-site mentor and supervisor for the apprentice.
- Provide work-based learning experiences in all aspects of the industry.
- Provide for the work-site rotations for all students.
- Provide part-time work opportunities for junior-year students during the school year.
- Provide full-time work opportunities during the eight-week summer period.

The participating employer must agree to provide a supervisor for each apprentice. These supervisors will need to be brought together by the program coordinator for training before being assigned an apprentice. The training will include the basics of the preapprenticeship program, goals and objectives of the program, and the specific kinds of activities that the apprentice will do in the particular job assignment. It should also include work behaviors to be instilled in the apprentice and lines of communications for problem resolution.

For School-Based Counselors and Teachers

Roles in a preapprenticeship should include:

- identifying students who can benefit from such preapprenticeship experiences
- screening, selecting, interviewing, and testing (if necessary) students
- meeting and discussing preapprenticeship with parents
- making recommendations of students to the partnership
- assigning students to appropriate academic courses to support the preapprenticeship program and the students' career goals
- arranging for tutors for those students needing additional support

All of these activities are originated and done by the counselors and teachers (and sometimes in consultation with the coordinator).

For the Community College Instructor/Coordinator

Roles and responsibilities include:

- college entrance counseling
- financial aid counseling
- college course articulation with high school faculty
- faculty visits to high school classrooms
- joint grant proposal development
- identifying college students to serve as mentors to preapprentices

For registered preapprenticeships, the state's department of labor representative will assume the role of the program coordinator. This person meets with the school-based personnel (counselor and instructor) and the parents of the student and goes over the procedures for apprenticeship. This person interfaces with the employer. In some locales, this person makes the initial contact with the employer; in other locales, the person works with the local high school to identify an appropriate placement.

Again, roles and responsibilities should be written down in a program manual. Each of the major tasks to be done (e.g., orienting parents, training work-site supervisors, and recruiting students) should have a standard procedure developed and written. This will probably be done by the program coordinator.

DEVELOP POLICIES AND PROCEDURES

The next step in the design of a preapprenticeship is to establish the standard operating guidelines for each of these major tasks. Let's look at each.

Recruiting Students

Student recruitment for a preapprenticeship program involves establishing minimum criteria for participation. A program in any area must start small in order for opportunities to develop as its worth is fully realized by the community. The Western Connecticut Medical Apprenticeship Project limited its participation to two students per school in the first year. Likewise, the Project ProTech program limited its initial enrollments. Student recruitment can be done singly by each of the participating schools or centrally by the partnership itself.

Establishing Specific Entrance Criteria

Entrance criteria must be established; for instance, in the Western Connecticut Medical Apprenticeship Project, the criteria are:

(1) Junior-level student
(2) A "C" average or better
(3) Completion of biology or chemistry
(4) Active participation by family
- Attend intake meeting.
- Sign student contract.
- Provide back-up transportation as needed.
- Provide appropriate work clothing.
- Arrange for immunizations and testing as required.

(5) Participation in all aspects of program orientation and training
(6) A commitment to stay with the program for two years
(7) Flexible schedule, which allows up to ten hours of work per week
(8) Student is motivated to explore medical careers. Priority is given to students who participate in medical careers clubs, have CPR certification, and/or have volunteer experience in the medical field.

As part of the written guidelines for the preapprenticeship program, these procedures should specify that students be targeted by the subject matter teachers in the various preapprenticeship program areas. This is one of the better places to identify students who can benefit from preapprenticeship programs. Likewise, the school guidance counselor should be formally included in the student identification process. The chain of referral should be from the teacher to the guidance counselor who then refers the student to the preapprenticeship program coordinator.

When the guidance counselor does identify a student interested in an apprenticeship, the counselor will determine the student's level of interest and career goals or aspirations and, where appropriate, discuss or administer an occupational inventory (e.g., ASVAB, Strong Vocational Interest Bank). The guidance counselor should also be charged with the responsibility of determining the student's eligibility criteria for the particular apprenticeship immediately upon meeting with the student to avoid problems later. The counselor should also arrange to meet with the student's parents to discuss preapprenticeship.

Interviewing

Apprenticeship committee interviews should then be held with students who meet the basic criteria as determined by the school guidance counselors and set forth by the program committee. Interviewers should be representative of all of the participating organizations. There should be a standard interview form used for all students to ensure fairness (see Figure 4.2). It is reasonable to set certain recruitment guidelines. For instance, in one of the

Interview Checklist

Student's Name ____________________ Parent's Names ______________

___ Explanation of the Program

___ Program Requirements

A. Specific to the Program
1. Immunizations
2. Transportation
3. Attendance at Program meetings
4. Meetings with Director and Mentor

B. Work Performance Issues
1. Dress Requirements
2. Job Attendance Requirements
3. Timeliness
4. Professional Behavior
5. Confidentiality and Safety Rules
6. Performance on assigned tasks

___ Performance Reviews

A. Formal Mid-Semester and End-of-Semester
B. Discussed with student & parents & counselor

___ Determining Student Interests

A. Strengths & Interests
B. Interest Batteries
C. Long-Term Work Goals

___ Review with Guidance Counselor

A. Approval of Work-Site
B. Class Scheduling

Figure 4.2 Interview checklist for preapprenticeship.

partnerships described herein, a minimum of two (2) students per school were selected for the first year of the program. Topics for an interview will probably include:

(1) Explanation of the program
(2) Program requirements (attendance, physical requirements, transportation, meetings)
(3) Work performance issues (e.g., dress code, professional behavior—need for confidentiality, performance of tasks, attendance)
(4) Performance reviews
(5) Student interest (e.g., likes and dislikes, career goals, other work experience)

Recruiting Employers in Each Community

Employer recruitment into new programs will require the assistance of senior officials of the partnership organization. Therefore, it is essential that people such as superintendents of schools and college presidents and even members of boards of education or trustees make calls and open doors for program coordinators. In the first year or so of a program, it might be necessary for the chief officer of the school and the program coordinator to meet with the business principals together to explain the program. After that, the program coordinator can probably do it alone.

Employers will want to know what the program is all about: who is involved, what the goals are, and the like. Have a fact sheet prepared, especially for an employer (see Figure 4.3). It should give all of the pertinent details, including what is expected of the employer if involved in the program. In the case of MechTech programs, local industry associations often serve as a focal point for employer recruitment.

Selection of employer work-sites is covered in detail in Chapter 7.

Orienting Students and Their Parents

Student and parent orientation has been identified as one of the most important aspects of a preapprenticeship program. In fact, it is reported to be the one single factor that determines overall success of preapprenticeship for students. It is no wonder that preapprenticeship sponsors will not even admit a student to a preapprenticeship unless there is active parental involvement in the process right from the start. Orientation should include the following kinds of information:

(1) A letter should be prepared and sent home to the parents (see Figure 4.4). It should end with a request to meet the parents face-to-face at a

How Can I Get More Information?

Contact the North Carolina Department of Labor at 1–800–LABOR-NC or the Department of Public Instruction, Vocational and Technical Education at (919)715–1665.

Contact the Director of Vocational and Technical Education in your local school system.

The North Carolina Department of Public Instruction
Bob Etheridge, State Superintendent

The North Carolina Department of Labor
Harry E. Payne Jr., Commissioner of Labor

The North Carolina Governor's Commission on Workforce Preparedness
James B. Hunt Jr., Governor

The State of North Carolina does not discriminate with regards to race, color, gender, national origin, or disability

Preparing for the Future

A JobReady Strategy

A Business and Industry Guide to High School/ Youth Apprenticeships in North Carolina

Figure 4.3 Employer fact sheet. (Reprinted with permission from N.C. Dept. of Labor.)

As technology and global competition increase the need for a well-educated and highly skilled workforce, business and industry in North Carolina will only be as strong as the quality of its' workforce. Ensuring that your company remains competitive requires investing in the quality of your employees. One way to do this is through the High School/Youth Apprenticeship Program.

High School/Youth Apprenticeship may be for your company if you want to:

- increase productivity and have highly skilled employees,
- participate in training of students before they enter the workforce, thereby reducing turnover,
- reduce the cost of training and retraining,
- attract adequate numbers of highly qualified job applicants,
- establish a communication link between business/industry and the education system.

What is a High School/Youth Apprenticeship Program?

- an industry-driven education and career training program based on recognized industry standards,
- a strategy to prepare youth to be high performance workers and lifelong learners,
- a means by which employers address current and projected employment needs through work-based and school-based learning,
- a partnership among business, industry, education, government, parents and youth apprentices.

Figure 4.3 (continued) Employer fact sheet. (Reprinted with permission from N.C. Dept. of Labor.)

How does a High School/Youth Apprenticeship operate?

- the apprentice enters a high school/youth apprenticeship in the junior or senior year in high school.
- the apprentice works on a part-time basis during the school year and full time in the summer.
- the apprentices' work-based learning is monitored and evaluated by the employer.
- related academic and technical instruction is coordinated by the school to connect work-based and school-based learning.
- after graduating from high school, the student moves into an adult apprenticeship program and continues his/her education, usually at the local technical/community college.
- when the apprentice successfully completes the required number of hours of work-based learning and related classroom instruction, certification of occupational and academic mastery is awarded.
- the apprentice has the option of entering the workforce and/or continuing his/her education.

Responsibilities of Participating Business/Industry

- participate in developing skill standards for the industry with the Department of Labor.
- work with the schools to create a comprehensive work process for the apprentice.
- interview and select potential high school/youth apprentices for your company.
- pay wages to the apprentice.
- provide an appropriate and safe work environment for the apprentice.
- assess the apprentice's progress and adapt work processes as necessary.
- provide a work-site mentor and supervisor for the apprentice.
- provide work-based learning experiences in all aspects of the industry.

Figure 4.3 (continued) Employer fact sheet. (Reprinted with permission from N.C. Dept. of Labor.)

Responsibilities of the Department of Labor

- develop a comprehensive, detailed work process that identifies industry standards and educational requirements.
- approve any waiver of identified child labor laws through the registration of the high school/youth apprenticeship program.
- assist industries in developing skill standards.
- monitor the program.
- issue certificate of occupational mastery upon completion of the adult program.

Responsibilities of the School

- assist with apprentice selection.
- provide a coordinator/school mentor who will give career counseling and guidance to the high school/youth apprentice.
- coordinate school-based and work-based learning.
- coordinate the apprenticeship program with the North Carolina Department of Labor and governmental agencies.
- provide flexibility for the apprentice in course scheduling.
- evaluate the apprenticeship program and award credit toward high school graduation.

Responsibilities of the Apprentice

- comply with guidelines established by the school and the employer for the apprenticeship program.
- commit to a long-term occupational and educational program.
- assume the role of worker and learner at the work-site.

Figure 4.3 (continued) Employer fact sheet. (Reprinted with permission from N.C. Dept. of Labor.)

ADAM High School
Danbury, Connecticut 06811-3443

October 21, 1995

Mr. and Mrs. Parents
Riding Trail Rd.
New Town, CT

Dear Mr. and Mrs. Parents:

Junior is being considered for the Town Superintendent's Medical Apprenticeship Program. This program has been made possible through the Workforce Development fund, area hospitals, local colleges and the region's high schools. This program has been designed to develop specific work skills that will make students employable in the health care fields and/or facilitate admission into post high school training and education.

There are three phases to this program. During the first semester students will explore health career options by rotating through a minimum of three health care settings. The following semesters will be spent at one worksite developing skills based upon long term employment and educational goals. Support for students will continue for one year after graduation either through mentoring, tutoring and/or follow-up with the staff.

I will work with Junior, you as his parents, and the employers and participating schools throughout the program. If approved, we will assist Junior in identifying the types of jobs he would like to explore and pursue. Once placements are arranged, a college mentor will be assigned to him to connect the work and classroom experiences. Along with job supervisors, I will monitor all students individually and as a group.

Junior's success in the program depends on your support. This is a two year program, requiring eight to ten hours a week. Depending on the worksite, these hours may be during the school day. Students must be able to complete the program and participate in all aspects of the training. As parents you are required to attend an Intake Interview with your child and sign the student contract. You must provide appropriate clothing, arrange for immunizations and testing as required, and provide back-up transportation as needed. Student progress reports will be sent to you on a regular basis.

Students are paid a minimum wage in their first semester and can earn salary increases in the following semesters based upon job performance. The enclosed information discusses student responsibilities. Please contact me with any additional questions or concerns. I look forward to meeting with you.

Sincerely,

Ruth F. Cantor, Program Director

Figure 4.4 Letter to parents.

specified time. If a large number of students are beginning a preapprenticeship at the same time, a group meeting might be appropriate, followed by individual parent meetings with the coordinator. Make sure that everyone knows why the program is happening and who the school's point of contact is for future correspondence and dialogue.

(2) The letter should identify the program, its goals and purpose. It should

indicate how long the program is and what kind of daily and weekly calendar it requires.

(3) The letter and follow-up meeting should provide parents with the overall procedures to be followed by the school in placing the student at the work site, who will supervise the student at the work-site, the kind of work the student will be doing, any dangers that the student might encounter, and the kind of credit and/or compensation the student will receive from the structured work experience.

(4) The parents' roles and responsibilities must also be discussed in person and in writing. These might include daily transportation to and from the work site, supplying of clothing or tools, insurance requirements, health screening for the student, or other specific items that might be of interest to or the primary responsibility of the parents.

Resolving Problems Arising on the Job

Almost certainly, when youngsters venture outside the confines of a school building, problems are going to arise, and in preapprenticeship programs, they can be pretty predictable. Inasmuch as the employer is in business to produce a product or service, the student's presence is secondary. Sometimes, students find this hard to deal with. After all, they are not center stage on the job. This oftentimes, initially, creates consternation for some students. Or the employer/supervisor may not be the right match for the student, and personality problems arise. Or the employer did not really understand what the entire program was about, and the student finds him/herself doing menial labor on a continuous basis, rather than what was agreed upon. The program coordinator should be ready for any or all of these kinds of problems to arise. If I can offer you one piece of general advice—have backup placement work sites ready and available. A standard grievance procedure document should be given to the student at the commencement of the placement to ensure that the student knows what to do if and when a problem arises. The grievance procedure should conform to the school's or college's procedure for handling student complaints.

Evaluating Students

Much like any other credit-bearing activity, work progress in cooperative preapprenticeship should be evaluated. After all, it is to carry academic credit weight; therefore, students should be made aware of the methods and criteria for evaluation. There should also be a formal procedure for evaluating students. Inasmuch as you will involve the field supervisor in this evaluation, you must take extra care to ensure that a procedure for weighing the student's on-the-job work is developed properly. More will be said about this in the chapter on evaluation (Chapter 9).

Hold at least weekly seminars for all apprentices to discuss work-based topics such as work behaviors, time sheets, work appearance, decision making, and so on. More will be said about this in the chapter on related classroom instruction (Chapter 8).

TRADITIONAL REGISTERED PREAPPRENTICESHIPS

A second category of secondary school preapprenticeship is the traditional preapprenticeship. Several state departments of labor have statutory authority to offer secondary school students a head start on traditional apprenticeship. For instance, the state of Connecticut in its statutes (Connecticut Regulations 31-51d 1-12) provides for preapprenticeship program registration.

The preapprenticeship program allows high school youth who are at least sixteen years old to start their apprentice training on a part-time basis while still in high school. The preapprentice registers with a program sponsor (employer) under the same conditions as an individual registering for a full-time apprenticeship; however, there is no provision for incremental wage increases, and, in some states, a preapprentice can be paid minimum wage. The preapprentice must also obtain parental consent to enter the program.

Hours the preapprentice accumulates after school, on weekends, and in the summer are credited hours toward completion of an apprenticeship should they choose to pursue that particular field after graduation from high school. The duration of preapprenticeship generally cannot exceed 2,000 hours in twenty-four months per regulation. Vocational-technical schools, such as the Connecticut system, often view this as an excellent form of cooperative work experience. It supplements the school's trade training experience with very good on-the-job opportunities. To administer this kind of program, a cooperative work experience agreement is developed (see Figure 4.5). This agreement should stipulate:

Student:

- The student is responsible for transportation to and from the job. The student must provide verification of automobile insurance if using a personal vehicle for such transportation.
- If the student does not report to school on a particular day the student shall not report to work.
- The student must obtain a work permit.
- The student agrees to be punctual, keep a job card up-to-date, return

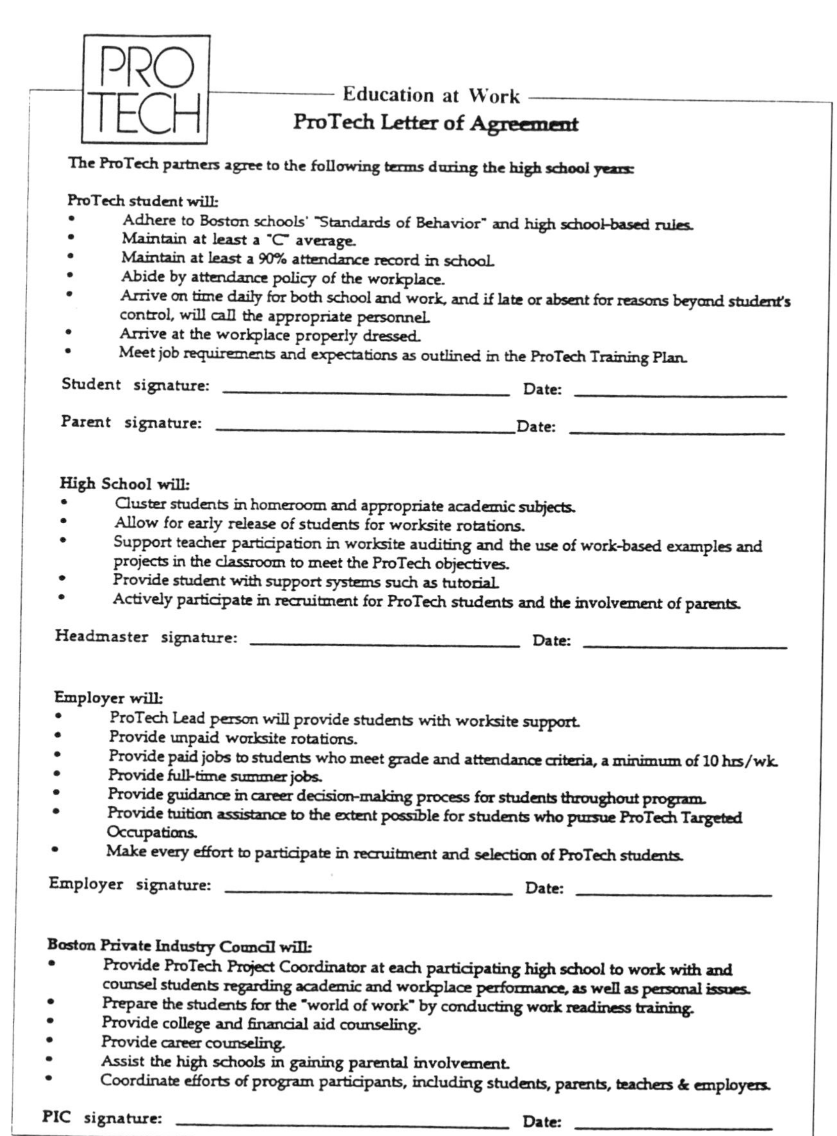

PRO TECH

Education at Work

ProTech Letter of Agreement

The ProTech partners agree to the following terms during the high school years:

ProTech student will:

- Adhere to Boston schools' "Standards of Behavior" and high school-based rules.
- Maintain at least a "C" average.
- Maintain at least a 90% attendance record in school.
- Abide by attendance policy of the workplace.
- Arrive on time daily for both school and work, and if late or absent for reasons beyond student's control, will call the appropriate personnel.
- Arrive at the workplace properly dressed.
- Meet job requirements and expectations as outlined in the ProTech Training Plan.

Student signature: ______________________ Date: ______________

Parent signature: ______________________ Date: ______________

High School will:

- Cluster students in homeroom and appropriate academic subjects.
- Allow for early release of students for worksite rotations.
- Support teacher participation in worksite auditing and the use of work-based examples and projects in the classroom to meet the ProTech objectives.
- Provide student with support systems such as tutorial.
- Actively participate in recruitment for ProTech students and the involvement of parents.

Headmaster signature: ______________________ Date: ______________

Employer will:

- ProTech Lead person will provide students with worksite support.
- Provide unpaid worksite rotations.
- Provide paid jobs to students who meet grade and attendance criteria, a minimum of 10 hrs/wk.
- Provide full-time summer jobs.
- Provide guidance in career decision-making process for students throughout program.
- Provide tuition assistance to the extent possible for students who pursue ProTech Targeted Occupations.
- Make every effort to participate in recruitment and selection of ProTech students.

Employer signature: ______________________ Date: ______________

Boston Private Industry Council will:

- Provide ProTech Project Coordinator at each participating high school to work with and counsel students regarding academic and workplace performance, as well as personal issues.
- Prepare the students for the "world of work" by conducting work readiness training.
- Provide college and financial aid counseling.
- Provide career counseling.
- Assist the high schools in gaining parental involvement.
- Coordinate efforts of program participants, including students, parents, teachers & employers.

PIC signature: ______________________ Date: ______________

Figure 4.5 ProTech cooperative/preapprenticeship agreement. (Reprinted with permission from the Boston Private Industry Council and ProTech.)

the card to the school instructor, and follow the rules and regulations of the employer.

School:

- The school department head will designate an individual to make periodic work-site visits.
- A representative of the state department of education may visit the job by appointment with the employer.
- In licensed occupations, the school must be notified that the student has been registered as a preapprentice through the state department of labor.

Employer:

- The employer participates in developing skill standards for the industry with the department of labor.
- The employer will work with the school to create a comprehensive work process for the apprentices.
- The employer participates in the recruitment and selection of students.
- The employer will interview and select potential high school/youth apprentices for the firm.
- The employer will pay wages to the apprentices in accordance with the state's wage and hour laws.
- The employer will provide an appropriate and safe work environment for the apprentice.
- The employer should assess the apprentices' progress and adapt work processes as necessary.
- The employer should provide a work-site mentor and supervisor for each apprentice.

The employer will not allow the student to work more than twenty-one hours per week while school is in session. Like all other apprenticeship programs, a specific job specification schedule outlining the requirements for the trade is to be followed by the employer in training the apprentice (see Chapter 5). This should be a part of the agreement for preapprenticeship work experience of high school age students (see Figures 4.6 and 4.7).

IMPLEMENTATION AND OPERATION OF PREAPPRENTICESHIP PROGRAMS

Certainly, a major concern in operating a preapprenticeship program is funding. Funding for many of the programs currently operating comes from

Cooperative Worksite Experience Program

PURPOSE

The mission of Connecticut's Vocational-Technical Schools is to provide intense occupational-specific training below the associate degree level. This equal opportunity program is offered for youth and adults in collaboration with business, labor, and community groups in preparation for the demands of the twenty-first century. Academic instruction is provided to ensure that students will be career-directed and will possess communication, problem-solving, and citizenship skills which allow them to be productive, adaptable, and satisfied individuals.

POLICY

A Cooperative Worksite Experience Program will be provided in the Regional Vocational-Technical Schools. The purpose of this program is to expand and enhance the student's learning with actual job site experiences, and to facilitate the transition from school to work. The program will be available to qualified students who have demonstrated readiness to benefit from a Cooperative Worksite Experience Program. A prerequisite will be compliance with all stated requirements and a completed Cooperative Worksite agreement between the student, parent/legal guardians, school, and the employer.

DESIGNATION OF AUTHORITY

The Superintendent of the Vocational-Technical School System is authorized to develop administrative procedures regarding all School to Career programs which includes the Cooperative Worksite Experience Program. The administrative responsibility for Cooperative Worksite Experience rests with the Vocational-Technical School Director.

The Superintendent shall provide for adequate staffing and funding to properly develop, run, and promote the program.

PROCEDURES

DRAFT

OBJECTIVES

The objectives of the Cooperative Worksite Experience are as follows:

To expand and enhance the student's learning through carefully planned unique career experiences in an actual work setting.

To help the student make the transition from school to work and career.

To teach the environment of work.

To increase the student's awareness and appreciation of the relevance of academic subjects as they apply to their occupational choice.

To create a holistic educational experience.

0637r48 -1- Last worked on March 1996

Figure 4.6 Connecticut cooperative education agreement. (Reprinted with permission, courtesy of Regional Vocational School System/Connecticut State Dept. of Education.)

Cooperative Worksite Experience Program

PROCEDURES

OBJECTIVES (Continued)

To provide the student with opportunities for potential career placement in their occupational choice.

To project a positive image for students through involvement in business and industry.

STUDENT ELIGIBILITY

The Cooperative Worksite Experience shall be available to any student of not less than sixteen (16) years of age who meets the criteria of the Cooperative Worksite Experience Program.

A Student may be admitted any time after the first marking period of the Junior year.

The student must have and must maintain a C or above in all subjects.

The student must demonstrate trade proficiency by being graded as acceptable or better on learning areas to which he/she has been exposed, based on performance criteria established in the curriculum.

For seniors commencing a cooperative worksite experience program before the end of the first marking period, eligibility will be based upon grades for the last marking period of the junior year.

The student must be in compliance with the school attendance policy.

A Student shall have written consent from their parents/guardian before they are enrolled in the program.

The student may be withdrawn from the program by request of the student's parent/legal guardian, the school, the employer, or the student.

Final decision of eligibility rests with the Director in consultation with the Department Head, Guidance Department, and School Coordinator (where applicable).

Any long-term day adult who has and maintains an average of C or above in shop and theory and has completed at least 85% of the prescribed adult program may be eligible.

DRAFT -2- Last worked on March 1996
0637r

Figure 4.6 (continued) Connecticut cooperative education agreement. (Reprinted with permission, courtesy of Regional Vocational School System/Connecticut State Dept. of Education.)

Cooperative Worksite Experience Program

HOURS OF EMPLOYMENT

While school is in session, Junior year students may be released up to but no more than, 10 hours a week. Senior year students may be released up to 20 hours a week as long as shop theory, BPR/etc. and graduation requirements are met.

Release time will be during the shop cycle only. Provisions shall be made to ensure that students in the program will have a minimum of ninety hours of shop theory per year, the fulfillment of time to be determined by individual units.

When school is in session, no student shall be assigned to a job which will require more than forty-five minutes of travel time one way, either to the job or to the student's home.

Final decisions, relative to hours of participation for all students rests with the Director in consultation with the Department Head, Guidance Department and School Coordinator (where applicable).

STUDENT RESPONSIBILITIES

The student, where applicable, will acquire an Employment Certificate (working papers).

The student shall agree to conform to the rules and regulations of the employer.

The student will keep a daily record (form provided) of the skills and jobs performed, and the record will be signed by the worksite mentor of the firm which employs the student. At the completion of each week, the student shall return this record to the shop instructor for evaluation.

If the student does not report to school when scheduled to do so, the student shall also not report to work. Any student scheduled to work, who will not be reporting to work must call his/her Department Head at the school before 8:00 A.M., and also notify their employer prior to the start of the work day.

The student shall be responsible for transportation to and from the job.

If driving to and from the job, the student shall provide and maintain a current driver's license and certificate of auto accident insurance, copies of both to be attached to a Cooperative Worksite Experience agreement.

0637r DRAFT -3- Last worked on March 1996

Figure 4.6 (continued) Connecticut cooperative education agreement. (Reprinted with permission, courtesy of Regional Vocational School System/Connecticut State Dept. of Education.)

Cooperative Worksite Experience Program

The student shall obtain written consent from his/her parent and/or legal guardian.

The student is responsible for maintaining grades and attendance as outlined in the Student Eligibility Requirements.

DRAFT

EMPLOYER RESPONSIBILITIES

An employer wishing to participate in Cooperative Worksite Experience will conform to all Federal, State of Connecticut Labor Laws and fair labor practices for which there are no special limitations such as license requirements. For a student to work in a licensed occupation (per latest Department of Labor List), he/she must be registered with the State of Connecticut as a preapprentice.

The employer shall provide a certificate of insurance to show that the student will have adequate insurance coverage while the student is employed and is on the employer's premises or assigned by the employer to a job off the premises. Adequate insurance means not less than on-the-job liability, workman's compensation, and auto if the student does any driving for the employer. Copies of these certificates should be issued to the school and must be renewed if the insurance expires before the end of the school year.

The employer agrees to instruct the student in safety procedures and safe work practices while involved in on-the-job training, and comply with all Federal, State and Local Laws.

The employer will notify the Department Head or Co-op Coordinator of any student absence as soon as possible on the day of absence.

Within 24 hours, the employer will notify the school of any accident or injury sustained by the student on the job.

The employer/monitor agrees to accept students, assign jobs and otherwise treat students without regard to race, color, religion, national origin, sex, sexual orientation, or handicap.

The employer will insure that no form of harassment, sexual or otherwise, is perpetrated upon the students.

The employer, in collaboration with the school, will chose a qualified employee as an instructor/mentor for the student for specific and related on-the-job training and career experiences. The assigned employee will provide instruction to the student in the areas of learning mutually established by the school and employer.

At the end of each work week, the worksite instructor/mentor will sign the Daily Work Record form indicating the student's training progress.

Wages paid to the student shall not be less than the Connecticut minimum wage.

0637r -4- Last worked on March 1996

Figure 4.6 (continued) Connecticut cooperative education agreement. (Reprinted with permission, courtesy of Regional Vocational School System/Connecticut State Dept. of Education.)

Cooperative Worksite Experience Program

SCHOOL RESPONSIBILITIES

The school will use the following guidelines to approve a cooperative worksite opportunity for students.

The proposed work experience will provide qualitatively unique career experiences to the student's course of study.

Worksite learning should be coordinated to the theory/trade being instructed in the school or as appropriate to the students learning plan.

The employer meets the training requirements and all the responsibilities as outlined in the Cooperative Worksite Experience Agreement.

The school Department Head, and/or Coordinator will work closely with the employer to achieve an understanding of the purpose and objectives of the Cooperative Worksite Experience Program, and to define the skills and training the student will be required to complete during the assigned period of employment.

The Director or his/her designee will review and verify all requirements and keep a central file with copies of all Cooperative Worksite Experience documents.

This program is designed to supplement a student's and the school's educational program. Nothing in the document shall be interpreted to mandate the offering of this program to all students.

0637r -5- Last worked on March 1996

Figure 4.6 (continued) Connecticut cooperative education agreement. (Reprinted with permission, courtesy of Regional Vocational School System/Connecticut State Dept. of Education.)

Cooperative Worksite Experience Program
Participation Agreement

AGREED TO BY:

STUDENT SIGNATURE DATE

DEPARTMENT HEAD SIGNATURE DATE

PARENT/LEGAL GUARDIAN SIGNATURE DATE

PHONE NUMBERS: HOME: WORK: EMERGENCY:

BEEPER FAX:

SCHOOL COORDINATOR'S SIGNATURE (As Applicable) DATE

COMPANY NAME ADDRESS

PHONE NUMBER(S) FAX NUMBER

AUTHORIZED EMPLOYER SIGNATURE TITLE DATE

SCHOOL DIRECTOR SIGNATURE DATE

All parties to this contract agree and warrant that in performance hereof no discrimination against any person or group of persons will be permitted on the grounds of race, color, religion, age, sex, sexual orientation, national origin or handicap in any manner prohibited by the laws of the United States or of the State of Connecticut.

0637r -6- March 1996

Figure 4.6 (continued) Connecticut cooperative education agreement. (Reprinted with permission, courtesy of Regional Vocational School System/Connecticut State Dept. of Education.)

CONNECTICUT STATE DEPARTMENT OF EDUCATION

COOPERATIVE WORKSITE EXPERIENCE PROGRAM PERFORMANCE OBJECTIVES

The ______________________ and ______________________ and
Employer School

______________________ agree to undertake a Cooperative Worksite Experi-
Student

ence Program for the purpose of providing work experience in ____________
Trade

beginning ______________, 19__ and ending ______________, 19__.

______________________ will be provided the opportunity to perform and
Student

develop, at a minimum, the following advanced skills or tasks:

1. ______________________
2. ______________________
3. ______________________
4. ______________________
5. ______________________
6. ______________________
7. ______________________
8. ______________________
9. ______________________

______________________ ____________
Department Head Date

______________________ ____________
Coordinator (as applicable) Date

______________________ ____________
Student Date

______________________ ____________
On-The-Job Instructor Date

0637r -7- Last worked on March 1996

Figure 4.6 (continued) Connecticut cooperative education agreement. (Reprinted with permission, courtesy of Regional Vocational School System/Connecticut State Dept. of Education.)

Cooperative Work Experience Program

Student Name:______________________ CO. Name:______________________

Term:______________________ Cycle:______________________

DAILY WORK RECORD

DATE	TIME (FROM-TO)	JOB WORKED ON (DESCRIPTION)	SUGGESTED SHOP/SKILL TRAINING (IF NEEDED)

INSTRUCTIONS: STUDENT IS TO KEEP DATE, TIME, AND JOB DESCRIPTION ON A DAILY BASIS. ON THE JOB INSTRUCTOR IS TO SIGN AND VERIFY, AND MAKE SUGGESTIONS AND COMMENTS AT THE END OF NINE DAY CYCLE.

COMMENTS: ______________________

ON JOB INSTRUCTOR'S SIGNATURE______________________ DATE______________________

0637r -8- Last worked on March 1996

Figure 4.6 (continued) Connecticut cooperative education agreement. (Reprinted with permission, courtesy of Regional Vocational School System/Connecticut State Dept. of Education.)

STUDENT PERFORMANCE EVALUATION

Name ______DRAFT__________________ Date __________

Department ______________ Instructor ______________

	U	F	S	G	E
JOB UNDERSTANDING Individual possesses a clear knowledge of the responsibilities and tasks he or she must perform.					
JOB PERFORMANCE The neatness, thoroughness and accuracy of the student's work.					
JOB PRODUCTIVITY The quality of the student's work in terms of volume and accomplishment.					
DEPENDABILITY Student can be relied upon in terms of being on time and completion of assigned tasks.					
COOPERATION The ability to work willingly with associates, subordinates, supervisors and others.					
ATTITUDE Student can be relied upon to act in a courteous and mature manner					
SAFETY/HOUSEKEEPING Practices safe, clean work habits on equipment and area of work.					

General comments as to student strengths, weaknesses and action needed to improve performance

____

____

____

____

Signature of Worksite Instructor/Mentor

E - Excellent F - Fair
G - Good U - Unsatisfactory
S - Satisfactory

Figure 4.6 (continued) Connecticut cooperative education agreement. (Reprinted with permission, courtesy of Regional Vocational School System/Connecticut State Dept. of Education.)

COOPERATIVE WORK EXPERIENCE PROGRAM

STUDENT NAME ____________________ CO. NAME ____________________

TERM __________ CYCLE __________

DAILY WORK RECORD

DATE	TIME: (FROM-TO)	JOB WORKED ON (DESCRIPTION)	SUGGESTED SHOP SKILL TRAINING IF NEEDED

INSTRUCTIONS: STUDENT IS TO KEEP DATE, TIME, AND JOB DESCRIPTION ON A DAILY BASIS. ON THE JOB INSTRUCTOR IS TO SIGN AND VERIFY, AND MAKE SUGGESTIONS AND COMMENTS AT END OF NINE DAY CYCLE

COMMENTS:

ON JOB INSTRUCTOR'S SIGNATURE ____________________ DATE ______

Figure 4.6 (continued) Connecticut cooperative education agreement. (Reprinted with permission, courtesy of Regional Vocational School System/Connecticut State Dept. of Education.)

North Carolina Department of Labor
Apprenticeship and Training Division
4 West Edenton Street
Raleigh, NC 27601-1092
(919) 733-7533

High School Apprenticeship Agreement

The program sponsor, the Industrial Training Coordinator for the affiliated high school, and the high school apprentice agree to the terms of the High School Apprenticeship Standards as part of this agreement. The sponsor will afford the high school apprentice equal opportunity in employment and training without discrimination because of race, color, religion, national origin, disability, or sex. Terms of agreement are on reverse side.

High School Sponsor: ______________
Address: ______________
____
Phone Number: ( ) ______________
Program # (File #) ______________
Supervisor of High School Apprentice: ______________

Employer (Sponsor): ______________
Is the employer the same as sponsor? ☐ Yes ☐ No
Name: ______________
Address: ______________
____
Phone Number: ( ) ______________

Social Security Number: ____/____/____
High School Apprentice: ______________
Address: ______________
____
____
Phone Number: ( ) ______________
Anticipated date for high school graduation: ______________
Date of graduation: ______________

Date of Birth ____/____/____ (M D Y)

Sex: ☐ Male ☐ Female

Race:
☐ White
☐ Black
☐ Hispanic
☐ American Indian
☐ Asian

Educational Background:
☐ High School Student
☐ Pursuing G.E.D.
☐ Other ______________
____
☐ Temporary Waiver
☐ Permanent Waiver

Trade Title
____
____

Dictionary of Occupational Titles (D.O.T. Code)
____

On-The-Job Training

	H.S. Plan	App. Plan
Hours Required	____	____
Credit for Previous Work Experience	____	____
Hours Remaining	____	____
Current Wage	____	

Related Instruction

	H.S. Plan	App. Plan
Hours Required	____	____
Credit for Previous Related Instruction	____	____
Hours Remaining	____	____

Wages Paid During Related Instruction: ☐ Yes ☐ No

Date High School Apprenticeship Begins ____/____/____ (M D Y)

Expected Date of Completion ____/____/____ (M D Y)

Attachments? (required if credit is given) ☐ Yes ☐ No

______________ Signature of High School Apprentice
______________ (Parent/Guardian, if a minor)
Date: ______________

______________ Signature of Employer (Sponsor)
______________ Signature of School Representative
Date ______________

______________ Signature of Apprenticeship Division Representative
Date ______________

Director's Approval:
______________ ☐ Previous Credit ______________ ☐ Waiver Date: ______________

White Raleigh Office
Green Apprentice
Canary Apprenticeship Representative
Pink Sponsor
Goldenrod High School

III 12 1945 3/95

Figure 4.7 North Carolina preapprenticeship agreement. (Reprinted with permission from the North Carolina Dept. of Labor.)

several sources: (1) Job Training Partnership Act funds and (2) School-to-Work Act Demonstration Grants, as in the case of the Project ProTech and the Cornell Youth Apprenticeship Project.

Funding is used to support full-time project coordinators, insurance, stipends for students (e.g., transportation, uniforms, tools), and other appropriate incidentals.

Other areas of concern in implementing a preapprenticeship program include:

(1) Work site selection: While selection of a work site will be similar for all apprenticeships and thus covered in Chapter 7, be extra cautious in placing preapprentices. Make sure the working environment is conducive to a youngster—in terms of safety, supervisor/mentor match, tasks to be carried out, and of course, location for travel purposes.

(2) On-the-job supervision: Close supervision is a must for the preapprentice. Thus, the supervisor/mentor must be just that. Work skill supervision is not enough for a youngster. The mentor must watch the other behaviors and concerns that will surface with a youngster. Make sure the mentor is aware of all those kinds of concerns we have with young people in a less structured environment away from school. More will be said about this in Chapter 8.

(3) Supervisor interfacing: Keep in close touch with the supervisor and make sure to make at least telephone contact once a week.

(4) Problem resolution: Deal with problems brought to your attention by either the supervisor or the preapprentice immediately. Do not let problems get out of control.

(5) Evaluation of apprentices: At the very beginning of the working relationship, go over the criteria for evaluation with the supervisor. Upon completion of the grading period or the end of the work assignment, go over the evaluation with the supervisor and make sure you understand the supervisor's positions on the various ranked items. More will be said about evaluation in greater detail in Chapter 9.

All of these topics will be discussed in detail in the following chapters on apprenticeship operations. Specific commentary will be made throughout the chapters for dealing with youth apprentices.

SUMMARY

We have discussed the growing area of preapprenticeship in this chapter. Understand that the preapprenticeship is much the same as the regular apprenticeship in many details, but different enough that a separate chapter is given to it. It is much like an appetizer if done right, in that it will motivate the student to want to continue in a structured "learn and earn" environment through the full apprenticeship. For further information, you may wish to contact:

Project ProTech
Boston Private Industry Council
2 Oliver Street
Boston, MA 02109 (617) 423-3755

Western Connecticut Superintendent's Medical Apprenticeship
Newtown High School
Newtown, CT (203) 426-7621

Jobs for the Future
1815 Massachusetts Avenue
Cambridge, MA 02140 (617) 661-3411

Project MechTech, Inc.
c/o Tedco, Inc.
70 Glen Road
Cranston, RI 02920 (401) 461-8605

Now that we understand how all of the pieces come together, let's get on to designing and developing apprenticeships!

REFERENCES

Boston Industry Private Industry Council. (1995). ProTech Orientation Materials and Promotional Materials. Boston, MA.

Federal Committee on Apprenticeship. (1992). *Youth Training and Education in America: The Role of Apprenticeship*. Washington, D.C.: Author.

Technical & Skills Training. (1993, May/June). "A Metalworking Youth Apprenticeship Program," pp. 3–4.

Technical & Skills Training. (1993, August/September). "Corning Has Upgraded Its Apprenticeship System . . .," pp. 34–35.

Western Connecticut Superintendents' Medical Apprenticeship Program. (1995). General Information Materials, Newtown, CT.

CHAPTER 5

Designing and Developing Apprenticeship Programs—Part 1

THE focus of this chapter is on designing and developing registered cooperative apprenticeships (see Figure 5.1). Now that it has been decided to train through apprenticeship, discussions between vocational-technical school instructors and administrators and/or college and cooperating firm's principals can begin.

Educational partners should understand that there are several ways in which registered apprenticeship programs are usually designed and developed. Depending upon a firm's structure and size and whether there is a collective bargaining agreement in place, an apprenticeship program can be designed and developed to be (1) delivered cooperatively with a labor union and state joint apprenticeship committee (JAC) or (2) delivered solely by the firm or agency or by an industry association representing several firms. In each case, the program can be registered with the department of labor. Each of these arrangements can and should be conducted cooperatively with a vocational-technical school and/or college. However, it should be understood that all these methods share some basic similarities in terms of processes for designing and developing programs of study. We will discuss each of these.

SIMILARITIES OF PROCESSES TO SET UP PROGRAMS

BASIC APPRENTICESHIP PROGRAM STANDARDS

The U.S. Department of Labor has established some basic program standards that provide a framework for apprenticeship program design (U.S. Department of Labor, 1989):

- sound and continuous employee-employer cooperation
- a minimum apprentice age of sixteen for apprentices

Figure 5.1 Apprenticeship design model.

- full and fair opportunity for all to apply for an apprenticeship
- selection of apprentices solely on the basis of qualifications
- a schedule of work processes in which the apprentice is to receive training and experience on the job
- organized instruction designed to provide the apprentice with knowledge in technical subjects related to the trade for a minimum of 144 hours per year
- proper supervision of on-the-job training with adequate facilities to train apprentices
- a progressively increasing schedule of wages
- periodic and fair evaluation of the apprentice's progress, both in job performance and related instruction
- the maintenance of appropriate records
- recognition of successful completion of apprenticeship

This chapter discusses each of the above basic components for apprenticeship program design.

BASIC COMPONENTS

Cooperative apprenticeship program design and development begins with determining a need for structured workplace training. Chapter 3 discussed needs assessment in detail. Basically, the following questions were addressed: Does there exist within a firm or industry:

(1) A high turnover in skilled areas?
(2) A problem in attracting skilled workers?
(3) New and changing technology?
(4) Program expansion within the next two to five years?
(5) Retirements among skilled workers within the next two to five years?
(6) A need for a more diverse skilled workforce?

If the answer is yes to any of the above questions, then cooperative registered apprenticeship could be the answer to a firm's personnel development needs.

IDENTIFY APPRENTICEABLE OCCUPATIONS

Apprenticeships can be developed in any area that can be adequately defined and that requires structured hands-on training. While there are many traditional occupations that are recognized for apprenticeship, the list continues to grow with new high-tech and high-demand occupations.

From the school's or college's perspective, identify occupations that are in demand by firms in your area. Make sure that firms seeking apprentices have a genuine need for workers in that field and a true desire and capability to train the apprentices for the occupation. Occupations need not be mechanical or construction-oriented; they can be in the health, computer technology, or office occupations; public services fields (e.g., fire and emergency medical services), or any other field that is well documented and in high demand. Appendix C is a list of apprenticeable occupations as published by the U.S. Department of Labor's *Occupational Outlook Handbook.*

ASSEMBLE THE ADVISORY COMMITTEE

In most structured programs involving organized labor, this committee is called a joint apprenticeship committee, or JAC. However nonlabor organized firms also use apprenticeship advisory committees for planning and development and to interface with state apprenticeship councils. Sound and continuous employee-employer cooperation is assured through communications via the advisory committee.

Design begins with organizing an apprenticeship advisory committee. The committee will work to accomplish several tasks related to apprentice program design, development, and operation. This committee will create linkages of education and business and industry (and to the entire community). The committee should "open doors" to potential firms that can benefit from apprentice workers. The committee should recruit potential apprentices, and the committee should initiate a signed agreement between apprentice and employer. The committee plans for and administers the related instruction. Together with the committee, you will identify appropriate occupational performance standards for the apprentice to perform in the given trade(s) or occupation(s). It is the focal point for coordination of the apprenticeship program by monitoring its standards and operation and documenting program completion. Once this is done, then apprentices and cooperating employers are recruited and matched together.

This committee oftentimes recruits apprentices, work sites, and journeymen to serve as instructors and arranges for related instruction through the vocational-technical school and/or community college. The committee assures equal access to all qualified applicants. Committee membership includes representation of:

- employers
- employer associations (key people who can sell the apprenticeship concept to members of their industry)
- journeymen
- apprentices

- Where the vocational-technical school and/or community college is directly involved in operating the apprenticeship program, then educational institution representation is also formally on the committee.

THEN, HOW APPRENTICESHIP TRAINING IS ORGANIZED

As was mentioned earlier, the design process begins with identifying and/or establishing an advisory committee. If your task is to use an existing apprenticeship committee as a basis for apprenticing your students, then an understanding of the apprenticeship committee structure is in order. Alternatively, if you are attempting to begin an apprenticeship program from scratch, perhaps in a nonunion situation, then establishing an advisory committee is in order. I will discuss both.

TEXTBOX 5.1

The Federal Committee on Apprenticeship

Statutorily, The Federal Committee on Apprenticeship (FCA) is one of the oldest permanent public advisory committees currently operating in the U.S. government. It was created by President Franklin D. Roosevelt in March of 1934 and continues under the National Apprenticeship Act of 1937 (29 USC 50). The committee is composed of twenty-five members appointed by the Secretary of Labor: ten represent employers; ten are from organized labor; and five, including the chair, are from the general public. The FCA advises the Secretary of Labor on apprenticeship and training policies and labor standards affecting apprenticeship and research needs. It meets twice yearly, and the meetings are open to the public.

Twenty-nine states have state apprenticeship agencies or organizations. Many of these state organizations have staffs to assist employers and/or unions and educational organizations develop and administer apprenticeship training. These state organizations also register apprentices for their programs and award a certificate upon successfully completing a program. The state apprenticeship office is also the contact for any funding available to help support the costs of apprentice training.

A LOCAL JOINT APPRENTICESHIP COMMITTEE (JAC)

Apprenticeship has had a long-standing tradition as a system of training designed and delivered cooperatively. As cooperative apprenticeship combines learning and earning, many different groups work together to coordinate program delivery. Joint apprenticeship committees, composed of representatives of management, labor, and professional associations, work together to develop and administer local apprenticeship programs. Businesses, trade associations, unions, and professional associations all come together to coordinate delivery of these programs, often with vocational-technical schools and/or community colleges. Often, the vocational-technical school and/or community college participates on the JAC (see Textbox 5.2).

TEXTBOX 5.2

The International Brotherhood of Electrical Workers

The International Brotherhood of Electrical Workers (IBEW) is one of the largest unions in the United States. It has had a long-standing interest in apprenticeship training. At the national level, the IBEW cooperates with the employer-contractors represented by the National Electrical Contractors Association (NECA). They form the National Joint Apprenticeship and Training Committee (NJATC) for the electrical construction industry. NJATC sets uniform standards for apprenticeship training, which provide program operation uniformity, and serves as a catalyst and resource for apprenticeship program operation. The IBEW has approximately ninety-two apprenticeship programs nationwide—many in cooperation with community colleges.

In the fall of 1977, the IBEW Local #3, Flushing, New York, ventured forward with an idea promoted by one of its most dynamic leaders, Harry Van Arsdale—to have new apprentices in the construction electricians division of the Local complete an Associate degree in labor studies as a required part of their apprenticeship training. This decision to add such a requirement and the accompanying effort to implement the educational innovation were unparalleled in the history of organized labor and apprenticeship training.

Local #3 and its employer-contractors represented by the NECA view a commitment to education as a well-established tradition in the labor movement. Additionally, the philosophical commitment to the apprenticeship program and to higher education and the Associate degree was so strong at that time that the union leadership established the degree program as mandatory in order to serve both the union's and industry's needs. Second, the local leadership viewed the need to raise the educational level of working people, thus improving their social status and self-esteem—and that of all workers as well. They foresaw payoffs for organized labor by impacting positively on the qualifications and characteristics of the bargaining unit, thus, they believed, strategically positioning labor for better union wage packages in collective bargaining.

Local #3's leadership believes that a union concerned about its own public image has a stake in educating both the public and its members about the labor movement. (Eighteen credits of the Associate degree program for apprentices is in labor studies.) Better educated apprentices make better union members—sympathetic to the cause: an argument touted for the program. In all, this college program is part of a total package, including technical skills training, a job in the trade, membership in the union and career mobility.

Your task, if a JAC is not to be involved in the participating industry, is to organize a cooperative apprenticeship advisory group. Much like any school's vocational or occupational advisory committee, it should be comprised of the firm's production management, the personnel manager, craft supervisors and master journeymen. In fact, the vocational-technical school craft and apprenticeship committees might be one and the same. To be officially recognized by the department of labor, however, the local apprenticeship committee will need to be structured as a joint apprenticeship committee. Again, a JAC will include representatives of local business and industry including all firms, unions if any, apprentice representatives, and educational agencies if involved in program delivery.

This committee will be responsible to do the following:

- Determine all of the knowledge and skills needed for the occupation(s) to be included in the program.
- Secure the cooperation of the workers and craft supervisors who will be expected to provide the apprentices with the direction and supervision on the job.
- Arrange for related instruction with the vocational-technical school and/or community college.
- Appoint the apprenticeship supervisor to maintain the program standards.
- Maintain all records.

JACs are involved in actual program operations. The JAC will oversee recruitment, program operation, apprentice evaluation, funding, and outside vendors such as the vocational-technical school and/or community college. I suggest that the educational institution's representative sit on this committee and participate in all decisions relating to curriculum, funding, staffing, and apprentice personnel matters. Textbox 5.3 describes one such organization's operations.

TEXTBOX 5.3

The California Professional Firefighters Apprenticeship Program

The California Firefighter Joint Apprenticeship Committee (CALJAC) is a nonprofit organization whose goal is to (1) provide training support for California's firefighters and (2) provide an avenue of entre into the fire service for qualified, underrepresented, and targeted workforce populations.

To participate in CALJAC, each fire department needs to make apprenticeship training available to all trainees. To be recognized as a bona fide CALJAC apprenticeship program, a fire department must describe in writing:

- how apprentices will be recruited and selected

- what training the apprentice will receive
- the length of the training period
- the wages to be paid the apprentice

Responsibilities for the program are shared by the California State Fire Marshall (CFM) (representing management) and the California Professional Firefighters (representing labor).

CALJAC specifies standards to control and ensure training program quality. CALJAC apprenticeship programs encompass seventeen different fire service occupations. As a three-year apprenticeship, the requirements for Firefighter I Certification is a total of 1,000 hours. Other CALJAC job titles include Firefighter Medic, Fire Engineer, Fire Marshall, and Fire Department Training Officer. To participate, each program must also provide all related and supplemental instruction; provide classrooms, instructors, drill grounds, A-V equipment, books, and supplies; and designate an instructor of record who holds a designated California subjects teaching credential and who oversees the program's operation. CALJAC also provides testing administration support services to each fire agency upon request. Upon completion of an apprenticeship program, a transcript is produced for a trainee, listing hours and a grade for completion, which can be presented to a local community college for credit consideration.

A typical fire service program receives CALJAC funding, approximately 10% of the total cost to support basic recruit training. Each participating fire department in California is eligible for reimbursement for apprentice training when its program is recognized as a CALJAC approved apprenticeship training program. Apprenticeship enables fire departments to share training resources. It offsets training costs to the fire department without changes in staffing, resource allocations, or company manning. In fact, recruits go on-line earlier and learn departmental-specific techniques and procedures.

ESTABLISH APPRENTICE QUALIFICATIONS AND SELECTION PROCESSES

Apprentice qualification is a statement of the qualifications and prerequisite skills the apprentice will need in order to qualify for apprenticeship training. These qualifications must be clear and objective and established by the apprenticeship advisory committee. Oftentimes, the vocational-technical school and/or community college staff and instructors will assist the committee by meeting with employers to come to terms on what actual qualifications a person needs to succeed in training for a particular occupation. Consideration of past experience and schooling and the requirements of the school or college programs for which the apprenticeship might be supporting should be factored into this decision as well. Qualification statements for apprenticeship programs should be consistent with the general requirements for Voc-Tech schools and/or community colleges if these are to be sources for the related training. However, it is reasonable that special requirements be stipulated if need be, such as the ability to be processed for a security clearance.

Consider this statement of qualifications as found in the printed materials of a variety of companies and organizations:

(1) Attain minimum age of eighteen years, with maximum age specified in some cases
(2) Have a high school diploma or G.E.D. certification
(3) Have earned units (a year-long study) in mathematics beyond arithmetic (algebra, geometry, trigonometry), in chemistry or physics, and in technology or computer science
(4) Can demonstrate a reasonable degree of mechanical aptitude
(5) Have good attendance and work performance records
(6) Pass a physical examination

Some firms stipulate that all employees of the company are eligible to participate in the firm's apprenticeship program; however, they must pass both a mechanical aptitude test and a mathematics and/or science entrance test. Furthermore, some firms also require that apprentices must complete up to four years of education in an accredited technical or vocational program, depending on the apprentice program. A process for recruitment, testing, and selection of apprentices needs to be specified. More will be said about this in Chapter 7.

DETERMINE NUMBERS OF APPRENTICES TO BE TRAINED

The federal standard specifies that the number of apprentices to be trained is usually a ratio of apprentices to skilled workers. It is seldom that a ratio of more than one apprentice to three skilled workers is feasible or effective. Such a ratio is based on the facilities available for employing and training apprentices and on future employment opportunities. Since apprentices learn from skilled workers, the quality of training largely depends on the number of skilled workers available to instruct apprentices and the ability of the skilled workers as instructors. The U.S. Navy shipyards, for example, has maintained a workforce of approximately 10 percent apprentices over the years. Employers must also consider the rates of retirement of their present workers and the skill levels of incoming workers. It is to the employer's advantage to use existing master craftsmen and technicians to train the younger worker who will be with the employer for years to come. For instance, the machine tool industry foresaw such a need, as pointed out in Textbox 5.4.

TEXTBOX 5.4

The Metalworking Connection

The Metalworking Connection is a manufacturing network, a nonprofit corporation. The Metalworking Connection initiated an 8,000 apprenticeship program for its sixty-seven metalworking firms in southwest Arkansas. This program is operated as a partnership of high schools and a community college, the U.S. Department of Labor, and the participating firms. The students are selected by the partnership and must possess the requisite math skills for the metalworking trades. Upon successful completion of this four-year program, the apprentice will be awarded a DOL Apprenticeship Certificate (journeyman standing) and have earned an Associate degree from the community college.

[Adapted from *Technical & Skills Training* (May/June, 1993), p. 4.]

IDENTIFY WORK PROCESSES

For a formal apprenticeship arrangement, each occupation should be defined in terms of its specific work processes (jobs). Standards for job performance must be identified and published.

Figure 5.2 is a work processes display (schedule of work experience) for a plumbing-heating-cooling mechanic from the state of Connecticut.

Where state or federal bureaus of apprenticeship do not have appropriate work processes schedules or when more detailed and up-to-date occupational information is required, many industry associations have such standards. For instance, the National Machining and Tooling Association has standards developed for machinists and related trades. The Office Systems Research Association has standards for occupations in the office areas. School-based personnel can assist the firm or industry by either identifying or developing work processes and standards for the occupation in question. Check the industry association representing the trade or occupation. A national standards project has been initiated to assist in compiling useful standards. Table 5.1 lists pertinent information about these standards. Any of these databases can be accessed by telephone modem by contacting

Training Technology Resource Center
Employment and Training Administration
U.S. Department of Labor
Washington, D.C. 20210
(800) 488-0901

CONNECTICUT STATE APPRENTICESHIP COUNCIL - CONNECTICUT STATE LABOR DEPARTMENT

PLUMBING - HEATING - COOLING APPRENTICE PRIME OBJECTIVE

___ Plumbing Mechanic *(P-2)*	4 years	862.381.030
___ Heating & Cooling Mechanic *(S-2)* *(Environmental Control Installer)*	4 years	637.261.014
___ Heating Mechanic *(Pipefitter) (S-4)* *(Environmental Control Installer) (S-8)*	4 years	637.261.014
___ Warm Air Heating & *(D-2)* Cooling Mechanic *(Environmental Control Installer)*	2 years	637.261.014
___ Cooling Mechanic *(L-6)*	2 years	637.381.010
___ Refrigeration Mechanic *(D-4)*	2 years	637.261.026
___ Oil Burner Servicer & *(B-2)* Installer *(B-4)*	2 years	862.281.018

Work Schedule "A"

The following schedule of work experience is intended as a guide. It need not be followed in any particular sequence, and it is understood that some adjustments may be necessary in the hours allotted for different work experience. In all cases, the apprentice is to receive sufficient experience to become fully competent and use good workmanship in all work processes which are a part of the trade. The apprentice will be fully instructed in safety and OSHA requirements. PLEASE NOTE: Connecticut Occupational Licensing Regulations require a minimum number of O.J.T. years *(calendar time)* as expressed below.

	(P-2) Plumbing Mechanic	*(S-2)* Heating & Cooling Mechanic	*(S-4)(S-8)* Heating Mechanic	*(D-2)* Warm Air Heating & Cooling Mechanic	*(D-6)* Cooling Mechanic	*(D-4)* Refrigeration Mechanic	*(B-2)(B-4)* Oil Burner Svcr/Instlr
A. Waste, Stacks, & Vents	1,900						
B. Water Piping *(Hot & Cold)*	1,200						
C. Setting of Fixtures	1,400						
D. Maintenance/Repair Pipes & Fixtures	1,200						
E. Pipe Cutting/Reaming/Threading to 2"	600	400	600	100	200	300	600
F. Process Piping-Misc. Gases/Liquids	600	300	200	100	200	200	
G. Tube Work	800	400	800	300	500	400	200
H. Commercial/Heavy Duty Piping (2 1/2"+)	200	500	600		100	300	
I. Welding, Brazing, & Joining	100	200	100				
J. Installation of Steam Heating Systems		1,300	1,800				
K. Installation of Hot Water Heating Systems		1,300	1,800				
L. Installation of Warm Air Systems		800		1,400			
M. Maintenance/Repair of Heating Systems		900	900	300			
N. Residential/Light Commercial Cooling Systems		600		800	1,300	500	
O. Commercial/Industrial Refrigeration		600		800	400	1,800	
P. Maintenance/Repair of Cooling Systems		200		200	1,300	500	
Q. Oil Burner Installation		400	900				1,500
R. Oil Burner Maintenance Repair		100	*00				1,700
TOTAL HOURS	8,000	8,000	8,000	4,000	4,000	4,000	4,000
S. Related Instruction Hours/Units	576	576	576	288	288	288	288
Minimum O.J.T Years	4	4	4	2	2	2	2

PLUMBING.WS

Figure 5.2 Work processes display for plumbing-heating-cooling mechanic—state of Connecticut. (Reprinted with permission from the Connecticut Dept. of Labor.)

Work Process Description

A. Waste, stacks, and vents:
1. Layout & installation from street sewer or septic tank to roof terminal.
2. Proper use & installation of adapters and fixtures, separators, grease traps, etc.
3. Preparing for and performing approved methods of testing pipe, fittings, & connections.

B. Water Piping (*Hot & Cold*)
1. Layout and installation from street service or well.
2. Proper sizing, use & installation of safety devices, fixtures, adapters to eliminate possible cross connections and hazards.
3. Preparing for and performing approved methods of testing piping and connections.

C. Setting of Fixtures
1. Installation of hangers, brackets for wall-hung fixtures.
2. Installation of kitchen, bathroom, sanitary fixtures including water coolers, heaters, drinking fountains, restaurant and medical equipment, etc.
3. Repair or replacement of faucets, mixing devices, etc.

D. Maintenance/Repair Pipes & Fixtures
1. Repair and replacement of pipe, fittings, hangers, etc. in drainage and water systems.
2. Proper, safe use of chemicals and power tools in cleaning drains & sewers.
3. Repair or replacement of faucets, mixing devices, etc.

E. Pipe Cutting/Reaming/Threading to 2"
1. Use and maintenance of hand tools
2. Proper, safe use and maintenance of power tools.

F. Process Piping-Misc. Gases/Liquids
1. Layout and installation of piping and fittings connecting pumps, machinery, compressors, tanks including by-pass, manifold and tandem schemes.

G. Tube Work
1. Tube Bending.
2. Mechanical assembly of pipe and fittings (flare - compression).
3. Brazing and soldering of pipe and fittings.
4. Handling, cleaning of tube and fittings for special applications.

H. Commercial/Heavy Duty Piping (2 1/2"+)
1. Layout and installation of hangers, pipe fittings.
2. Set-Up, safe use and maintenance of power tools.
3. Safe handling and storage of material.

I. Welding, Brazing, & Joining
1. Proper, safe use and care of equipment.
2. Joining of pipe and fittings using gas or electric welding.
3. Cutting material ox-acetylene.

J. Installation of Steam Heating Systems (*High and Low Pressure*).
1. Layout and installation of boiler, fittings, pipes, controls, and radiators.
2. Testing system for leaks, proper operation using appropriate tools, gauges, and instruments.

WAGE SCHEDULE
0 to 1000 hours________
1001 to 2000 hours________
2001 to 3000 hours________
3001 to 4000 hours________
4001 to 5000 hours________
5001 to 6000 hours________
6001 to 7000 hours________
7001 to 8000 hours________
Journey Person Rate________

Figure 5.2 (continued) Work processes display for plumbing-heating-cooling mechanic—state of Connecticut. (Reprinted with permission from the Connecticut Dept. of Labor.)

K. Installation of Hot Water Heating Systems

1. Layout and installation of boiler, fittings, pipes, controls, and radiators.
2. Testing system for leaks, proper operation using appropriate tools, gauges, and instruments.

L. Installation of Warm Air Systems

1. Layout and installation of furnace, controls, filters, and humidifiers.
2. Testing system for proper balance and operation using appropriate tools, gauges, and instruments.

M. Maintenance/Repair of Heating Systems *(troubleshooting)*

1. Repair or replacement of boiler, fittings, controls, safety devices, etc.
2. Purging or flushing system.
3. Testing system for proper operation using appropriate tools, gauges, and instruments.

N. Residential/Light Commercial Cooling Systems

1. Layout and installation of compressors, motors, evaporators, and condensers in separate, packaged, or sealed units.
2. Installation of gaskets, insulation, and accessories.
3. Charging systems with proper gas.
4. Testing system for leaks, proper operation using appropriate tools, vacuum & pressure gauges, and required instruments.

O. Commercial/Industrial Refrigeration

1. Layout and installation of walk-in coolers, freezer cases, central air conditioning systems, industrial cooling and freezing systems.
2. See "N", Items 1 through 4.
3. Installation of refrigerant metering, pressure, temperature, and motor controls.
4. Installation of coil-forced air, gravity, multiple-flood, and two-temperature systems.

P. Maintenance/Repair of Cooling Systems *(troubleshooting)*

1. Repair or replacement of compressors, motors, evaporators, condensers, control lines, dryer/filters, valves, insulation, gaskets, and accessories.
2. Purging, bleeding, removing air and moisture from systems.
3. Recharging with proper gases after repairs.
4. Testing system for leaks, proper operation using appropriate tools, vacuum & pressure gauges, and required instruments.

Q. Oil Burner Installation

1. Layout and installation of burners, combustion chambers, oil storage tanks, controls and accessories.
2. Purging, adjusting, testing burner for efficient combustion, safe and proper control operation using appropriate tools, gauges, and instruments.
3. Checking power plant through all phases of operation for proper safety device operation.

R. Maintenance/Repair of Oil Burners *(troubleshooting)*

1. Repair or replacement of burners, tanks, chambers, motors, controls, etc.
2. Proper use of instruments and procedures to diagnosis breakdowns and malfunctions.
3. Preventive maintenance procedures for efficient operation
4. Checking power plant throughout all phases of operation for proper safety device operation.

NOTE: All installation and repair must be made strictly in accordance with manufacturer's specifications, State Occupational Licensing requirements, codes, and OSHA requirements under supervision of properly-licensed journey people and contractors.

Figure 5.2 (continued) Work processes display for plumbing-heating-cooling mechanic—state of Connecticut. (Reprinted with permission from the Connecticut Dept. of Labor.)

TABLE 5.1. **Representative Labor Skill Standards Projects.**

Project Title	Point of Contact	Description	Trade Area	Source
Skill Stds. for Electronics Ind.	C. Fields-Tyler (408) 987-4289	Skill stds. for key electronics jobs	Mfg. speclst. Adm/Info svc. Pre/post sales	Am. Electronics Assn. 5201 Great American Pkway. Santa Clara, CA 95056
Skill Stds. for Health Sci. and Tech	Sri Ananda (415) 241-2712	Skill stds. for healthcare core	Therapeutic Diagnostic Info. Systems Environmental	Far West Lab for Ed. Rsch. & Dev. 730 Harrison St. San Francisco, CA 94107
Laborers-AGC Ind. Standards Prog.	John Tippie (203) 974-0800	Construction project laborer jobs	Pipe laying Concrete work Lead remed. Petro-chem.	Laborers-AGC Ed. & Trng. Fund 37 Deerfield Rd. Pomfret Ctr., CT 06359
Voluntary Industry Stds. for CPI Technical Workers	Kenneth Chapman (202) 872-8734	Chemical process jobs	Entry-level chemical tech. Process tech.	American Chemical Society 1155 15th St. NW Washington, D.C. 20016
Skill Stds. for Hospital and Tourism Industry	Doug Adair (202) 331-5990	Front-line jobs in hospitality and tourism	Food server Cashier Front desk clerk	Council on Hotel, Restaurant and Institutional Ed. 1200 17th St. NW Washington, D.C. 20036
Skill Stds. for Electronics Industry	Irwin Kaplan (202) 955-5810	Entry-level electronics techs.	Maintenance tech.	Electronics Industry Foundation 919 18th Street Washington, D.C. 20006

TABLE 5.1. (continued).

Project Title	Point of Contact	Description	Trade Area	Source
Skill Stds. for Computer Aided Drafting	John Morrison (202) 662-8905	CADD techs.	CADD tech.	Foundation for Industrial Modernization 1331 Pennsylvania Ave. Washington, D.C. 20004
Skill Stds. for Advanced High Peformance Mfg. Jobs	C. J. Sholl (202 662-8968	Core skills for all techs.	Core skills needed for high perf. technology	Foundation for Industrial Modernization 1331 Pennsylvania Ave. Washington, D.C. 20004
Skill Standards for the Retail Industry	Robert Hall (202) 783-7971	Retail sales associate skills	Sales clerk	National Retail Federation 710 Pennsylvania Ave. Washington, D.C. 20001
Skill Stds. for Industrial Laundry Industry	Geoffrey Northey (202) 938-5057	Entry-level production and maintenance techs.	Production tech. Maintenance tech.	Uniform and Textile Service Assn. 1730 M Street NW Washington, D.C. 20036
Skill Stds. for Entry Level Welders	Nelson Wall (305) 443-9353	Entry-level welding skills	Basic welder	American Welding Assn 550 NW LeJeune Rd. Miami, FL 33126

THE PROCESS OF OCCUPATIONAL ANALYSIS

Work processes are developed after either identifying, accessing, or conducting an occupational (job/task) analysis for the occupation. The occupational analysis should be done by observing qualified incumbents performing on the job and by reviewing documentation (e.g., standards projects, written procedures, technical manuals) associated with the occupation. Occupational analysis is a systematic process of defining an entire job or job category (e.g., an electronics technician). Task analysis is a process by which you can better understand the task and how it should be trained and evaluated. When conducting a task analysis, you should specify the exact behaviors your apprentice will have to exhibit to perform the task.

Task Analysis

Task analysis is the process of breaking down the task into pieces. An entire occupational analysis can be thought of as a tree. The thick branches at the top are called *jobs* (work processes). Jobs are the major segments within a field; in law enforcement these might be (a) crime prevention, (b) patrol, (c) emergency services, (d) detection, and (e) community education (Cantor, 1992).

The second level (thinner branches) are called *duties.* Duties are the logical subsets of jobs. This is often about the amount of material that would be covered in a single training lesson. For example, some units under "patrol" might be (a) traffic laws and procedures, (b) traffic control, (c) controlled access highway patrol, and (d) deploying personnel and equipment.

The third level (even thinner branches) are called *tasks.* Tasks are the logical subsets of duties. They are often about the size of a single instructional objective within a lesson. For example, some tasks under "deploying personnel and equipment" in a police science course might be (a) dispatch unit, (b) set up equipment, (c) provide backup, and (d) provide other support agents.

The fourth level (thinner yet branches) are called *task elements.* Task elements are the logical subsets of tasks. They are often about the size of a supporting or enabling instructional objective in a course. For example, some task elements under "provide backup" might be (a) select backup sources and (b) establish communications.

The fifth level (thinnest branches) are called *steps* (under task elements). Steps are the operations required to accomplish tasks or jobs. For example, some steps under "establish communications" might be (a) turn on two-way radio, (b) shift to transmission frequency, (c) push mike button and (d) speak clearly, and so on.

It is sometimes difficult to determine how much detail should be used in breaking the task apart. For example, would it be better to break the "turn on two-way radio" into two steps as follows? (1) Grasp the switch with thumb and forefinger; (2) move the switch to the ON position. Or would it be better to combine them into one set as follows? (1) Throw the switch to the ON position. The answer is that it depends on your apprentices. The general recommendation is that apprentices with lesser prerequisite skills (e.g., preapprentices) will need a finer detailed analysis with more steps (smaller bites). Write the task analysis with steps of a size that seems reasonable to you. Then, if experience in teaching the task shows the task should be broken into smaller steps, make additional substeps in your task analysis. Thus, different task analyses written for the same task may include more or fewer steps, depending on the learning history of your apprentice. A sample job and task analysis is diagrammed in Figure 5.3. The outcome of making such a detailed diagram allows you to map the identified training need and begin to design the instruction. It shows where your apprentices' entry-level skills actually are and, accordingly, where training is needed and should begin.

A task analysis will also help you to design an on-the-job teaching strategy and diagnose any instructional problems encountered during or after instruction. I have found that it also helps to flowchart the tasks for competence in preparation, presentation, application, and evaluation. In preparation, a good task analysis will help you to decide the best way to demonstrate the work process to your apprentices. It helps you to fully know and understand the task. Often, with multiple steps in a given task, you may decide that it would be inappropriate to lecture through all the steps and then demonstrate all of the steps in a single demonstration. You may decide to instruct the first six steps and then evaluate apprentice mastery to this point. You will make sure that they master those six steps before going on to subsequent steps. The task analysis helped you take a reasonable approach that will pay off in terms of quicker acquisition by your apprentices and less frustration for you as well. A task analysis will also prove helpful in your evaluation of your apprentices' progress towards mastery.

WRITING AND USING INSTRUCTIONAL OBJECTIVES

An instructional objective is a statement of exactly who the apprentice is, what the apprentice must perform, under what circumstances or conditions, and to what degree or standard of proficiency (Cantor, 1992). Again, an instructional objective provides the communication link between you, the on-the-job instructor, and your apprentices. After receiving instruction, the only way that you can know if training has been effective is when an apprentice can perform in the desired manner.

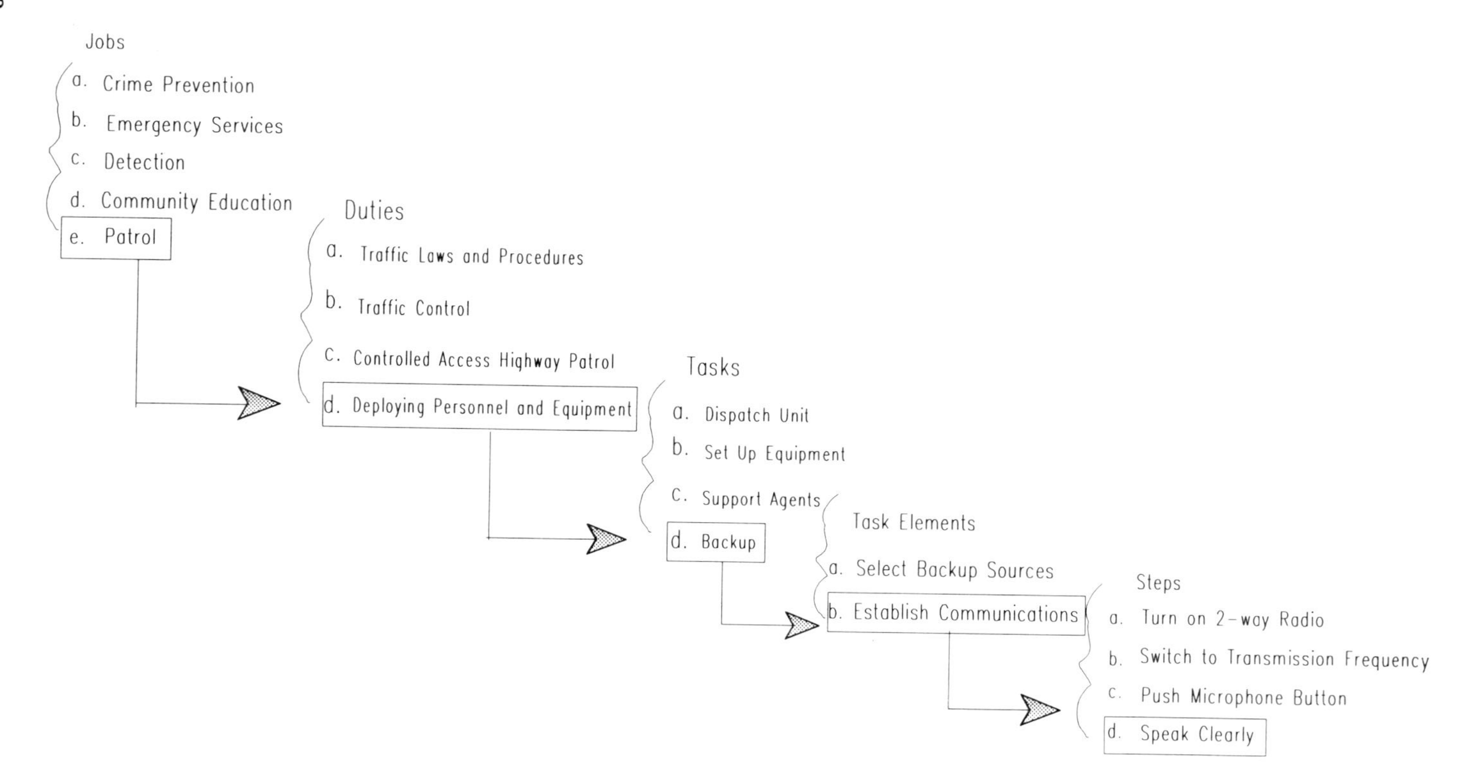

Figure 5.3 Sample job/task analysis.

Once the job and task analysis is completed, decisions about what information should be formally instructed must be made. This requires answers to questions such as: What must your apprentice do immediately upon entering on the job? What can be learned less formally as the job is performed by your apprentice? What is critical to job performance regardless of when your apprentice will be called upon to perform the task? Once these kinds of questions are answered, you are ready to write your instructional objectives.

Format of an Instructional Objective

A well-written objective has a standard format. This format includes four parts known as (source: National Fire Academy):

- audience
- behavior
- condition
- degree (or standard)

When writing your instructional objective, you must make sure that the person reading the objective can readily answer four questions:

- Exactly who are the intended apprentice learners? (audience)
- Exactly what do you want the apprentice to do? (behavior)
- What conditions and/or limitations will the apprentices be operating under when they do what you want them to do? (condition)
- What are the indicators of the apprentice's successful accomplishment of the objective? (degree or standard)

In the following sections, audience, behavior, condition, and degree (or standard) will be defined and discussed in detail.

Instructional objectives are developed to accurately describe job or task performance. The task of writing conditions and standards for each and every instructional objective is often very difficult, but well worth the effort.

Audience

"The *audience* represents who the apprentice is (e.g., the police officer; the firefighter; etc.)" (National Fire Academy). Information that is needed to develop good instruction includes learner reading ability, past training, job rating, special needs, age, and so on.

Behavior

The definition of behavior is stated as follows: "The *behavior* part of an instructional objective indicates what the apprentice must do, produce, or demonstrate in order to demonstrate achievement of the instructional objective" (National Fire Academy). Clarity of the behavior part of the objective is extremely important. The apprentice and the instructor must have the same idea or definition in mind concerning what must be done.

When you write an instructional objective, you must specify an action verb. The action verb you choose will be in the "behavior" part of the objective and must be observable and measurable. You must make a decision as to which kind of behavior is indicated and use the appropriate verb. The action verb must always match the kind of learning capability desired. As you can see, it would be quite difficult to measure, given the words appreciate, understand, internalize, and so on. How would you know if someone had learned? On the other hand, the "doing" words are much better in showing what is expected in the "behavior" area.

Write your instructional objective's behavior in a form that will enable your apprentices to show the learned behaviors you wish them to exhibit. For example, if you wish your apprentices to recognize all the parts of a certain valve, write the objective as follows:

> Given a picture containing twenty-five (25) parts from various valves, circle *only* the parts pertaining to Brand X gate valves. *All* Brand X parts must be circled within five (5) minutes.

In this instance, your desire is that your apprentices must memorize all the parts of Brand X gate valves, recognize them by sight, know how many parts there are, and differentiate these parts from other valve parts not of Brand X-type. If your apprentices can successfully complete the objective as written, you can be certain they have the required skills and knowledge.

A performance of a behavior can be detected by one or more of the five senses. If you can outwardly detect and judge an apprentice's performance by a simple physical observation, that performance would, in essence, be the test. If the performance were to bake an apple pie, the learner's performance would be the finished product. The pie could be judged by appearance, taste, smell, texture, weight, and shape. No "indicators" would be needed to demonstrate hidden performance. With good performance objectives written, you will now be able to specify your work processes in measurable terms and develop good lesson plans and evaluation measures. Lesson plans will be described in Chapter 8.

Separate On-the-Job and Related Instruction Performance Objectives

Once the kind of behavior to be performed by the apprentice is identified, then it is important to separate those performances that will be taught and mastered on the job from those that will be taught and mastered in the classroom. Ensure that performance objectives that call for hands-on performance (e.g., turn a screwthread, tie a bowline knot) are identified for work-site training. Likewise, performance objectives that indicate cognitive mastery (e.g., calculate pump pressure, describe fire behavior) must be assigned to related classroom activities.

ESTABLISH TERM OF APPRENTICESHIP

The term of apprenticeship is determined at the time of the design of the apprenticeship program. It is determined based on the sum total of the time to be devoted to mastery of each work process plus the amount of time to be spent in related classroom instruction.

ALLOCATION OF WORK TRAINING TIME

A primary feature of cooperative apprenticeship is that it is designed precisely to occupational specifications; therefore, adequate time is incorporated into the program to prepare workers for each work process—no more and no less! As part of the occupational analysis of the trade or occupation, it is important to determine both the time required to perform and the relative difficulty and importance of each job or task (work process). Then in the design of the apprentice's training program, you can allocate the amount of training time required to master each aspect of the job.

Program design includes making determinations about what activities will happen on the job. For example, for a plumbing-heating-cooling mechanic's apprenticeship training to be most effective, the apprentice must be engaged in meaningful activities all of the time he or she is on the job. That is why a rotational concept is often engineered into the design. Generally, to make this happen, the apprenticeship agreement spells out allocation of work training time in actual clock hours. To make this determination, you must use the data derived from the occupational analysis above. First, determine the number of work processes (jobs). List these as illustrated:

Work Process

Process A—Waste, stacks, and vents
Process B—Water piping
Process C—Setting fixtures
Process D—Maint/repair pipe/fixture
Process E—Pipe cutting
Process F—Process piping
Process G—Tube work
Process H—Comm. piping
Process I—Welding
Process J—Install steam
Process K—Install hot water
Process L—Install warm air

Then decide on the relative difficulty in performing each. The thought here is that more difficult jobs will take longer to master. Use a Likert-Type Scale (1–5) to rate these (e.g., 1 is easiest and 5 is most difficult). In theory, therefore, it would take five times as long to master a process ranked 5 as it would a process ranked 1. Place the numerical indicator next to each work process.

Work Process

Process A—1	Process G—3
Process B—2	Process H—2
Process C—2	Process I—3
Process D—4	Process J—1
Process E—2	Process K—1
Process F—1	Process L—3

Next, determine how long it will take (in hours) to train on the shortest work process. Then, multiply the number of hours it will take to train on a work process by the numerical indicator of difficulty already assigned to that work process, as shown below:

Work Process

Process A—320	Process G—960
Process B—640	Process H—640
Process C—640	Process I—960
Process D—1280	Process J—320
Process E—640	Process K—320
Process F—320	Process L—960

Finally, sum up the clock hours on the work process listing.

Total = 8,000

It is important to consider term of apprenticeship, because it indicates the level of technical difficulty and complexity of the occupation and requisite instruction. It also is an indicator of the magnitude to which the vocational-technical school and/or community college will need to be involved in cooperatively delivering the program. A typical statement of term of apprenticeship is

> The term of apprenticeship will be 8,000 working hours and the successful completion of all academics that are scheduled to end prior to July of the apprentice's fourth year.

Sometimes a firm will permit credit for previous work experience or training. A typical statement for credit for previous experience is:

> At the discretion of the Joint Apprenticeship Committee, up to 1,000 hours of credit may be given to successful applicants for apprenticeship. After discussion with a candidate's prospective manager, the Joint Apprenticeship Committee may award credit based on experience in the machine shop department, a related production trade at this firm, or other similar employment. Any credit will be applied toward the work process portion of apprenticeship. If such credit is granted, alternate work experience shall be provided as a substitute at the discretion of the Joint Apprenticeship Committee. There will be no academic credit awarded.

PLAN RELATED CLASSROOM INSTRUCTION

The background knowledge required to be proficient on the job also needs to be specified for apprenticeship as indicated earlier. While the basic federal standard herein only calls for 144 clock hours per year, undoubtedly far more than this minimum will be specified. One maritime firm's stipulation is as follows:

> Apprentices shall be required to attend weekly classes in related classroom instruction for a total of five hundred (500) hours for the full term of apprenticeship and at such other times as may be required by the Trade Training Instructor. Hours spent in related instruction classes shall not be considered as hours of work, but shall apply toward completion of apprenticeship.

One public service agency has the following stipulation in its document:

> In addition to on-the-job training, each apprentice shall be required to satisfactorily complete a minimum of one-hundred forty-four (144) hours of related vocational instruction for each year of apprenticeship. The curriculum shall be approved by the Committee. The Committee will review attendance and progress reports of each apprentice. Time spent in this related instruction will not be considered as hours of work.
>
> The Committee will approve the school, training program, and hours of instruction the apprentice will use to obtain the related training.
>
> If the Committee designates that the related training be from schools which require the payment of tuition fees, the agency through the existing tuition refund program, will reimburse the student for the cost of such tuition, upon satisfactory completion of the course. Other miscellaneous charges such as books and supplies are the responsibility of the apprentice.

Community colleges offer related education in different ways, from customized noncredit classes to actual courses towards college degrees.

A typical course program display at American River College in California is (1992–1993 Catalog, p. 87)

Sheet Metal Service Technician

150A First Course in Apprenticeship (2.5)
Introduction of Sheet Metal Service Technician Apprenticeship—orientation, tools and components, basic electricity, and basic electricity heat controls, air filters, blowers and fans.

150B Second Course in Apprenticeship (2.5)
Covers basic controls, heat-cool combination, electrical theory and motor design, construction and operation. Diagnosis and repair of refrigeration lines.

151A Third Course in Apprenticeship (2.5)
Review electrical theory; practical electrical review; air distribution system, residential and commercial; pneumatic controls.

151B Fourth Course in Apprenticeship (2.5)
Basic piping, and heat pump circuitry and controls, expansion valve systems, capillary tubes, chimneys, vents and flues.

152A Fifth Course in Apprenticeship (2.5)
Service person time management, residential oil-fired warm air furnace, industrial and commercial refrigeration systems.

152B Sixth Course in Apprenticeship (2.5)
Covers trouble shooting of electrical furnaces, electronic filters, gas and oil furnaces, cooling systems and heat pumps.

153A Seventh Course in Apprenticeship (2.5)
Covers boilers, chilled water systems, air distribution and air balance.

153B Eighth Course in Apprenticeship (2.5)
Pneumatic and electrical control systems, hydraulic and control. Energy load management systems.

Vocational-technical school- and community college-related education support will be covered in detail in Chapter 8.

PLAN SUPERVISION OF APPRENTICES

Apprentices are customarily under the immediate supervision of a senior-level skilled worker to whom they have been assigned. This person may be an individual who has been trained as an instructor for apprentices or may simply be the employer. In any event, a person must be specified as the "master" skilled worker for the sake of the apprenticeship. Statements relating to supervision of apprenticeship programs include the following:

> The responsibility for the administration and coordination of the apprentice program will be assigned to the chairman of the committee. In the

event of disagreement between the agency and the apprentice on disputes arising concerning provisions of the local apprenticeship standards, either party may consult the registration agency for an interpretation of opinion. Both parties may also avail themselves of the grievance procedures of the existing collective bargaining agreement.

ESTABLISH APPRENTICE WAGE SCHEDULE

A common method of expressing the apprentice wage, or for that matter arriving at a wage progression schedule, is a percentage of the skilled worker's rate in a particular occupational title. A progressively increasing wage schedule should include increases every six months. Successful apprenticeships recognize the need to motivate the worker. Good work processes demonstrated through evaluations merit wage increases. The increases should be scheduled throughout the apprenticeship to provide both a monetary incentive and reward for steady progress on the job. As the term of apprenticeship ends, the apprentice should be approaching 90% of the rate paid to the skilled worker at entry level in the occupation. A usual statement found in an apprenticeship standard is: "The apprentice wage progression shall be subject to mutual agreement between the employer and the union." Figure 5.4 is an apprentice wage and progression chart used by the state of Connecticut for recording wage progression agreements between the employer, apprentice, and state DOL.

EVALUATION OF APPRENTICES

A process for regular evaluation of apprentices, in conjunction with the progressive wage schedule, also needs to be determined. Evaluation will be treated in depth in Chapter 9.

REVIEW APPRENTICE AGREEMENT

The program must provide for signing an agreement of apprenticeship between the apprentice and the employer. The agreement states that the apprentice will work for the employer for a specified period of time and that the employer will pay a specified wage and provide the apprentice with specified skills and competencies.

The agreement should contain:

- name, home address, and birthdate of the apprentice
- name of the employer
- term of apprenticeship

AT-6/24 (Rev. 05/95)

APPRENTICE WAGE PROGRESSION CHART

Trade #1 ______________________ Period (Month, Hours, Years) #1 ________ (No) (Period)

#2 ______________________ #2 ________ (No) (Period)

#3 ______________________ #3 ________ (No) (Period)

Effective date ____________ Work week ________ Hours

	Time Intervals		Minimum of		Minimum of		Minimum of
1st	____ to ____	#1	____	#2	____	#3	____
2nd	____ to ____		____		____		____
3rd	____ to ____		____		____		____
4th	____ to ____		____		____		____
5th	____ to ____		____		____		____
6th	____ to ____		____		____		____
7th	____ to ____		____		____		____
8th	____ to ____		____		____		____
9th	____ to ____		____		____		____
10th	____ to ____		____		____		____
11th	____ to ____		____		____		____
12th	____ to ____		____		____		____
Current Minimum Completion Rate:*		$	____/Hour	$	____/Hour$	$	____/Hour

Note: Pursuant to U.S.C.F.R. Title 29, Part 5 and C.G.S. 31-53, apprentices assigned to Prevailing Wage Project jobs must be paid their percentage of the program sponsor minimum completion rate or the project journeyperson's rate (Prevailing Wage), whichever is higher, plus 100% of the fringe benefits listed in the wage determination for their occupational classification.

Firm: ______________________ Address/Zip Code ______________________

______________________ Signature of Company Official ______________________ Printed Name and Title ________ Date Signed

*This Apprentice Wage Progression Chart will be updated periodically by the sponsor.

Journeyperson License and Social Security Number

Name	Social Security Number	License Type & Number

______________________ Sponsor Name and Address and Zip Code

______________________ Signature of Company Official ______________________ Printed Name and Title ________ Date Signed

The Commissioner of Labor's Work Training Standards for Apprenticeship and Training Programs (Regulations of Connecticut State Agencies, Sec. 31-51d (1-12) and Chapter 393 of the Connecticut General Statutes require direct supervision of apprentices. The Regulations of Connecticut State Agencies, Sec. 20-332-15a (c) specify that "No apprentice shall at any time engage in any of the work for which a license is required without direct supervision. Direct supervision shall mean under the guidance of a licensed contractor or journeyperson...." The signer attests that this sponsor will adhere to the apprentice supervision requirements and that all journeypersons listed above are employed full-time on the job site and are available to train the sponsor's apprentices.

Figure 5.4 Apprentice wage and progression chart. (Reprinted with permission from the Connecticut Dept. of Labor.)

UNITED STATES DEPARTMENT OF LABOR
BUREAU OF APPRENTICESHIP AND TRAINING
60 PARK PLACE - ROOM 339
NEWARK, NEW JERSEY 07102

☐ New Program ☐ Vet
☐ Registration ☐ Non Vet
☐ Revision ☐ Student-Learner

STATE OF NEW JERSEY
DEPARTMENT OF EDUCATION
DIVISION OF VOCATIONAL EDUCATION
CN 500
225 WEST STATE STREET
TRENTON, NEW JERSEY 08625

ATR ☐ PROGRAM ☐

APPRENTICESHIP STANDARDS/APPRENTICESHIP AGREEMENT JOINT APPROVAL

WORK PROCESSES MUST BE ATTACHED AS REQUIRED

PRIVACY ACT STATEMENT: The information requested herein is used for apprenticeship program statistical purposes and may not be otherwise disclosed without the express permission of the undersigned apprentice. Privacy Act of 1974—L.L. 93-579. NJAC 6-3-20.

* 1. Social Security No. — — 2. Name of Apprentice LAST FIRST
3. Street Address 4. City 5. State 6. Zip Code
7. App. Municipal Code 8. Sex (M/F) 9. Ethnic Group W=White B=Black AA=Amer. Indian/Alaskan AP=Asian/Pacific Isle H=Hispanic Y=Yes N=No 10. Telephone — —
11. Vet. Status Y/N 12. Military Def. Status Y/N 13. Econ. Disadvantaged Y/N 14. Date of Birth Month Day Year / /
15. D.O.T. Code 16. Trade Occ. Title
17. Date Registered 18. Federal Registration No.

I have read and understand the conditions of the apprenticeship standards/apprenticeship agreement.

The apprenticeship standards referred to herein are hereby incorporated in and made a part of this agreement.

Signature Date

20. Union Affiliated Y/N 21. Sponsor Employer Name 22. County Code
23. Street Address 24. City
25. State 26. Zip Code 27. Telephone Number 28. Total Employees
29. Empl. Municipal Code 30. S.I.C. Code Industry Type 31. Length of Program (Months) 32. Date Apprenticeship Began
** 33. Prior Credit Employment (Months) ** 34. Prior Credit Related Training (Hours) 35. Public Vocational School Related Training Y/N

36. Provision for related instructions ____

37. Number of Journeyworkers in this Trade ____ Ratio: 1 Apprentice to ____ Journeyworkers Job Site ☐ Work Force ☐ Department ☐ Plant ☐
Authorizing ____ apprentices.

38. **Wage Schedule** (Construction apprentice wages must be expressed in percentage of journeyworker rate.)

1st period $ ____ per ____ (% of Journeyworker rate ____) 5th period $ ____ per ____ (% of Journeyworker rate ____)
2nd period $ ____ per ____ (% of Journeyworker rate ____) 6th period $ ____ per ____ (% of Journeyworker rate ____)
3rd period $ ____ per ____ (% of Journeyworker rate ____) 7th period $ ____ per ____ (% of Journeyworker rate ____)
4th period $ ____ per ____ (% of Journeyworker rate ____) 8th period $ ____ per ____ (% of Journeyworker rate ____)

Based on journeyworker rate of $ ____ per ____ for a standard work week of ____ hours. Rate of Overtime ____ Probationary period ____ months

I have read and understand the conditions of the apprenticeship standards/apprenticeship agreement.

39. ____ Name and Title of Sponsor's Authorized Official (Print) ____ Signature ____ Date

40. ____ If Bargaining Agency (Print Name, Local Number) ____ Signature ____ Date

Federal Representative ____ Date

Registered with the Bureau of Apprenticeship and Training, United States Department of Labor, as incorporating the basic standards recommended by the Federal Committee on Apprenticeship.

Registered By ____ Date

Bureau of Apprenticeship and Training
United States Department of Labor

State Representative ____ Date

Approved by the New Jersey Department of Education, Division of Vocational Education as incorporating the basic standards set forth in the State Plan for Vocational Education.

Approved By ____ Date

New Jersey Department of Education
Division of Vocational Education

Distr. Sponsor ☐ Apprentice ☐ Coord ☐ State ☐ BAT-Nwk ☐ BAT-Fd ☐ BAT-Reg ☐ Linkage ☐

STATE USE 41. Apprentice Approver Co L E A 42. School Training Site Co L E A 43. Coop Program Co L E A 44. Cooperative Linkage Y/N

45. Co-op Teacher Coordinator (Print or Type) ____ Date ____

*Print or type all information in black ink for photocopy purposes
**A written explanation must be attached for all prior credit

Figure 5.5 Apprentice indenture agreement (U.S. DOL—NJ).

- wage schedule
- length of probationary period
- an outline of work processes schedule, number of hours per year the apprentice agrees to attend classes, subjects, and name of participating school
- any other special provisions such as credit for past work experience
- signatures of employer and apprentice

Where required, a union endorsement/JAC approval process can also be incorporated into the agreement. Figure 5.5 is a sample copy of the U.S. Department of Labor's form for apprenticeship indenture agreement.

SUMMARY

Chapter 5 has presented the essential apprenticeship program components. Chapter 6 will now focus on developing the requisite documentation to underwrite apprenticeship program operation, including development of an apprenticeship program standard and indenturing agreements.

REFERENCES

Bath Iron Works Corporation. (1988). *Standards of Apprenticeship.* Bath, ME.

Bath Iron Works Corporation. (1988). *Standards of Designer Apprenticeship.* Bath, ME.

Cantor, J. A. (1992). *Delivering Instruction to Adult Learners.* Toronto, Canada: Wall & Emerson.

Electrical Joint Apprenticeship and Training Committee. (1989). *Local Apprenticeship and Training Standards for the Electrical Contracting Industry.* New York: Author.

Empire State College. (Undated). *Apprentice Handbook.* New York, NY: Author.

MechTech, Inc. (1990). MechTech, Inc., Promotional and Descriptive Materials. Greenville, RI.

New Jersey Turnpike Authority. (1985). New Jersey Turnpike Authority Apprentice Agreement. New Brunswick, New Jersey.

New York State Department of Labor, Apprenticeship and Training Council. (1987). *An Overview of Apprentice Training.* Albany, NY: Author.

State of Wisconsin, Bureau of Apprenticeship Standards. (1991). *A Guide to Apprenticeship in Wisconsin.* Madison, WI: Author.

U.S. Department of Labor. (1989). *Setting up an Apprenticeship Program.* Washington, D.C.: Bureau of Apprenticeship and Training.

Designing and Developing Apprenticeships—Part 2: How Cooperative Apprenticeship Programs Are Developed

OVERVIEW

IN Chapter 5, each of the components of a cooperative apprenticeship were discussed. Now we need to take the planning done in the design phase and actually develop apprenticeship opportunities for our students (see Figure 6.1). In order to do this, let's quickly review what we know about the various organizations and participants needed to make a cooperative apprenticeship work.

One very important role that an educator serves in apprenticeship program development is that of a catalyst for partnership formation with business and industry for the implementation of apprenticeships for training. The vocational-technical school and/or community college must serve as more than the institution providing related education for an apprenticeship. It now must serve as a catalyst, bringing together all of the necessary partners to deliver training. How this can happen is the focus of part of this chapter.

Cooperative apprenticeship involves the participation of firms needing skilled workers (and sometimes industry associations), labor unions, journeymen workers willing to mentor and train apprentices (oftentimes labor unions), the apprentices themselves, educational institutions, and departments of labor (see Textbox 6.1, which describes a machine tool firm's JAC). Let's look at the various forms that cooperative apprenticeship can take. Specifically, they are as follows.

TEXTBOX 6.1

Joint Apprenticeship Council Functions

The functions of the JAC respective to the apprenticeship training program include the following:

- Set minimum standards for training and education in all occupations

Figure 6.1 Apprenticeship program development.

under its aegis. These standards are now to be derived from National Tooling and Machining Association (NTMA) industry standards.

- Set minimum experience requirements for apprentices. It was decided, based upon review of the NTMA standards work processes, that this would be a four-year or 8,000-hour apprenticeship.
- Establish procedural guidelines for training program operation. These include testing and selection of apprentices, employer selection, working hours, wage progression, and so on. These will be discussed.
- Appoint a coordinator for apprenticeship program operations. This person's salary is shared by the foundation and college.
- Ensure that adequate supervision is available for all apprentices on the job. The coordinator will provide supervisor training for all employer-supervisors selected for the program.
- Hear and make decisions respective to grievances of apprentices.
- Establish testing procedures that meet the standards of fairness and equal opportunity for apprentice recruitment and retention, as well as final certification as journeymen. Equal opportunity is essential in apprenticeship programs.
- Establish procedures for granting certificates of completion through the state or federal DOL to apprentices who complete all of the program requirements.
- Ensure that all facilities and equipment are available for training. The coordinator will work with the college personnel to ensure that both the work sites and college facilities meet the requirements of the trade.
- Review requests for credit for previous experience.

---

1. Many apprenticeship programs are union-centered, wherein the workers' collective bargaining organization and the firm's management cooperate in the operation of the apprenticeship program. The local joint apprenticeship committee is comprised of members of management (the employer), the union, journeymen and apprentices, and possibly the educational institution. The training program that the local JAC will follow is usually provided by the union's national JAC (see Textbox 6.2). The vocational-technical school's and/or community college's efforts should be directed at working with the local JAC to support the apprenticeships. To do this, ensure that a JAC representative from those unions that you seek to work with is a formal member of your college advisory committee(s). More is given about this in Chapter 8.

TEXTBOX 6.2

The International Union of Operating Engineers

The International Union of Operating Engineers (IUOE), representing heavy equipment operating engineers in the construction industries, supports dual-enrollment degree programs nationwide. In the late 1970s, IUOE initiated fifty-three Associate degree programs in twenty states, at that time involving 2,300 apprentices.

IUOE leadership believes that a parallel exists between the technical knowledge and educational discipline required of an indentured apprentice in the operating engineer trades and the knowledge required of an undergraduate student in the first two years of study. Like most high-technology industries, rapid technological changes in equipment and building processes necessitate better trained personnel in the operating engineer trades. IUOE does view these programs as a means to help operating engineers keep up with the rapid changes taking place in the trades by working cooperatively with management and public education. They also view the college degree as a means to achieve better trained management personnel in the future.

The basic funding for a program operated by an IUOE labor union also comes from the 1 percent journeyman wage tax levied upon the employer. The average IUOE journeyman's wage is $28 hour, which generates approximately $4.5 million annually for training. U.S. Code (29 CFR 29) stipulates that 62 percent of an average journeyman wage be paid to registered apprentices. The contractor's payroll tax levy for apprenticeship training varies from IUOE Local to Local.

Community colleges nationwide support IUOE programs; IUOE programs vary according to the local college curriculum but tend to emphasize technology. Catonsville Community College has a full time college apprentice program coordinator to interface with all unions in program delivery. Catonsville services IUOE and other union programs in an off-site facility dedicated to these kinds of programs. Other colleges such as Fox Valley Technical College support the development of apprenticeship instructors through instructor training at universities such as at the University of Wisconsin at Stout. The IUOE Locals use national standards to develop and implement their programs. Each Local has curriculum autonomy. IUOE college programs vary in means for award of credit for on-the-job training. The Maryland college programs have a stipulated written agreement for eighteen credits, or one-third of the degree program. A brochure describes the process for earning credit to the apprentice degree candidate early into the program. Arizona also appears to have this kind of process for eleven college credits; other programs use college-internal challenge for credit processes (WA), ACE/LEAP process (RI), and less formally structured processes elsewhere. IUOE programs are three years in length. If construction is slow and there is no work for apprentices for a period of time, apprentices must work approximately 80 percent of that year to advance. Students in both of these programs must attend on their own time, and they are not paid to attend.

---

2. Some apprenticeship programs are strictly single-firm operated. The firm does all of the operational work, and the apprenticeship program committee is comprised of management, journeymen, apprentice representatives, and training personnel. To effectively work with small firms, you will need to help them identify apprenticeable occupations and locate students wishing to be trained. Strategies for this will be discussed in Chapter 7. Alternatively, you can work with firms in specific occupational categories (e.g., bakeries, machine shops) and help them organize into educational

foundations with dedicated JACs and then provide assistance to them in developing and delivering apprenticeships. More about how you can work to establish an educational foundation to promote apprenticeships will follow in this chapter.

3. Still other apprenticeship programs are industry association or large industry operated. Here, the program is designed to train in a specific family of occupations, all related to one specific industry. The industry association might have already established an educational foundation [501(c)(3) type of corporation] to administratively control the apprenticeship program. The JAC is comprised of members of each of the participating firms, the industry association, journeymen, apprentices, and the educational organization handling the related education and training (see Textbox 6.3). The vocational-technical schools and/or community college's role is to support the educational foundation with those services best provided by an educational institution, including student recruitment, testing and screening, related classroom training, and counseling. If a foundation has not been established, you might suggest this as one mechanism to catalytically support program development.

TEXTBOX 6.3

Nonprofit Educational Foundations

Organizations that have traditionally been involved in delivery of occupational and career education have increasingly turned attention to the nonprofit educational foundation as a form of organization for program delivery. In its basic structure, the nonprofit organization is simply a legal corporation formed under the provisions of the Internal Revenue Service's statutes. In this manner, the board of directors of such a corporation can permit most of the revenues generated by the organization to support the actual training and development activities.

A very attractive feature of an educational foundation is that the various participants in the educational activity can come together under one umbrella organization independent of any of the participants' member organizations. In essence, all of the participants are on equal ground, a very necessary feature that was discussed in Chapter 2.

Nonprofit educational foundations have been used in various industries with great success. For instance, the Tidewater Maritime Training Institute (TMTI) at Norfolk, Virginia, was initiated to provide preapprenticeship training in the maritime trades. TMTI was formed in 1979 by fourteen shipyard employers in the Tidewater area. A professional trade association was established under IRS Code 501(c)(6), and a tax-exempt educational foundation, named the Maritime Trades Foundation, was created under IRS Code 501(c)(3). A concomitant board of directors representing the companies governed both groups. The board of directors hired an executive director to assume leadership of the institute.

A key feature of this program is the collective collaboration of the employers electing to participate in the foundation. Corporate resources, both in-kind and monetary, are lent to the foundation for day-to-day operations. This has included $1 per year lease of a facility for training, loans of journeymen as instructors, work-site mentoring opportunities, and tools and materials. Additionally, each participating firm knows that it has a responsibility to make full-time employment available for program graduates.

This program is but one of many such collaborative efforts using an educational foundation as a catalyst for an educational partnership. Similar stories exist in the automotive, public service, construction, and finance industries.

---

4. Some apprenticeship programs are administratively centered in a community college. Here, the college supports the administrative details of the apprenticeship program and the recruitment of apprentices. It then recruits participating firms and places the apprentice. The JAC is a central or regional committee representing several prominent trades (see Textbox 6.4). The vocational-technical school or community college will oversee all of the aspects of the apprenticeship program operation with the assistance of a state department of labor and multiple state and local JACs (see Figure 6.2).

TEXTBOX 6.4

Northcentral Technical College Apprenticeship Programs

Northcentral Technical College (NTC) provides apprenticeship programs and program support for fourteen different trade programs ranging from barber/stylist to machinist. NTC operates an apprenticeship office directed by the associate dean. The office provides a central place for information and coordination of apprenticeship programs.

Trainees desiring apprenticeship training may visit NTC's apprentice office to receive general information about the apprenticeship training process. The trainee is then referred to a list of potential employers who consider sponsoring an apprentice. Alternatively, trainees may secure employer sponsorship first and then proceed to the college for formal registration. Once an apprentice and employer are matched, then NTC supports the formal indenturing process. This indenturing agreement establishes the framework for the training.

A joint apprenticeship advisory committee receives an apprentice's application and reviews it for acceptance into a program. The JAC monitors an apprentice's progress over the duration of the program.

NTC provides formal related courses for all of the programs. These classes are on-campus and often in the evening. Journeyman craftspeople work as adjunct instructors for the courses.

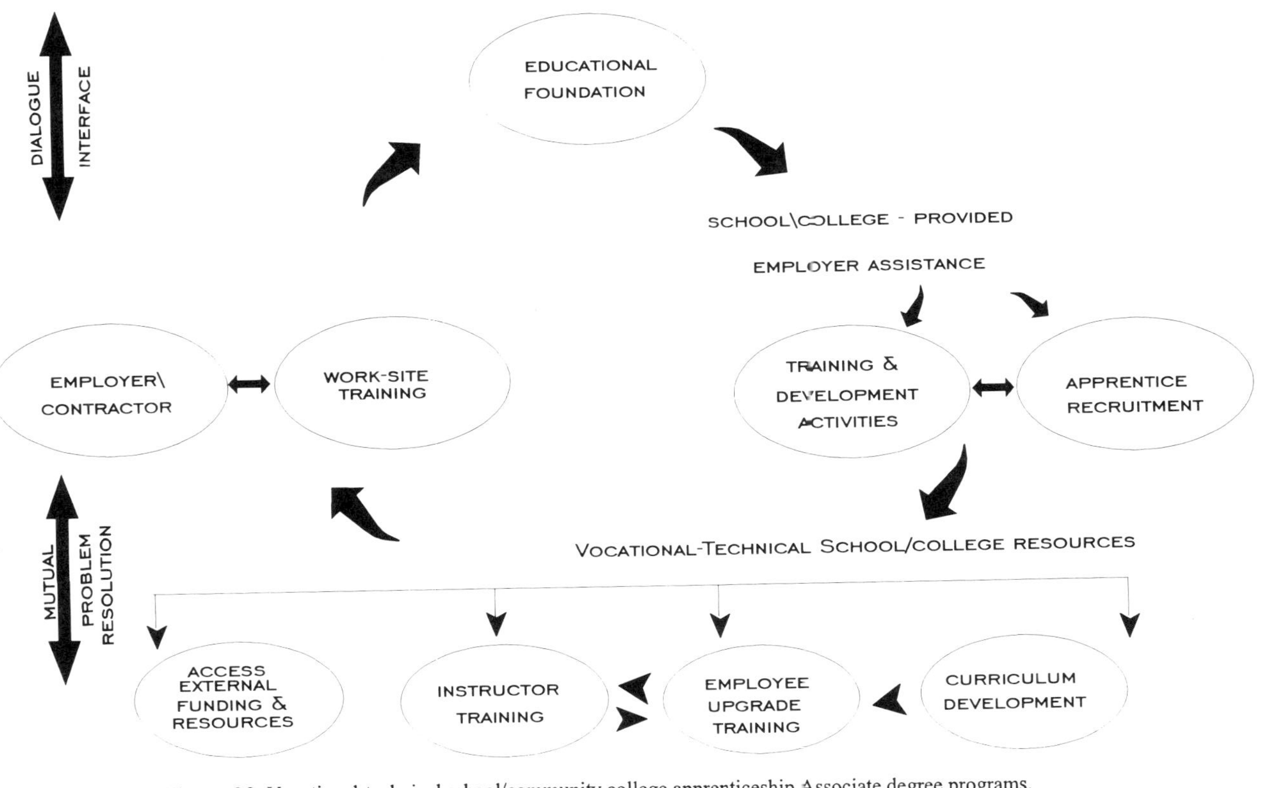

Figure 6.2 Vocational-technical school/community college apprenticeship Associate degree programs.

TABLE 6.1. **Apprenticeship Configurations.**

Where There Is a . . .	Then Work with a . . .	And Education's Role
National or state joint apprenticeship committee (JAC)	Local JAC	Coordinate and facilitate
National trade association	Local educational foundation	Coordinate and facilitate
National manufacturer	Local manufacturer group or association	Coordinate and facilitate
Small stratified group of businesses	Training foundation by occupational group	Coordinate and facilitate

5. Lastly, some apprenticeship programs are centered in the state government department of labor. Here, through regional offices, the state does the recruiting of apprentices, usually through its employment and training offices and/or vocational-technical schools. The state assists participating firms in registering the apprentice and then in monitoring progress. The vocational-technical school provides the related training or assists the firm to provide same. Its role becomes one of providing related education (see Textbox 6.5). Table 6.1 lists these various configurations and educational roles.

TEXTBOX 6.5

The Connecticut Department of Labor's Apprenticeship Programs

In 1937, Congress passed the Fitzgerald Act, named after Congressman William Fitzgerald of Norwich, Connecticut, who sponsored the legislation. He was a state senator, congressman from the second district, deputy labor commissioner, and director of the Connecticut State Employment Service.

The Connecticut State Apprenticeship Council and the apprenticeship system was established in 1938 by executive order. The council's main function is to set standards and promote the establishment of training programs in industry. The council was set by statute in 1959. The council is composed of twelve members who, in their everyday job, are to be involved in some way with apprenticeship training. Four members are from labor, four from industry management, and four from the public sector. One of the public members is the deputy commissioner, who also serves as chair. They advise the labor commissioner on apprenticeship matters and formulate policies for the effective administration of apprenticeship. Connecticut Apprenticeship Training Division, located within the Connecticut Department of Labor, was established in May 1946, to assist the council in the administration of apprenticeship.

Apprenticeship in Connecticut is regulated by the Commissioner's Work

Training Standards for Apprenticeship and Training (Connecticut Regulations 31-51d - 12) based on the Code of Federal Regulations 29 part 29 and Equal Opportunity in Apprenticeship and Training (Connecticut Regulations 46a-68 1 - 17) based on Code of Federal Regulations 29 part 30.

People desiring to learn by apprenticeship in Connecticut may enter a program in one of several ways. The state department of labor recruits apprentices through its nine area job service centers. Bureau of apprenticeship representatives will assist aspiring apprentices to select an apprenticeable occupation and identify a firm or business agreeing to sign an agreement to train the person in a formal apprenticeship. The representative will then oversee the relationship for its duration. Alternatively, the state bureau of apprenticeship representative will work with businesses seeking apprenticeable trainees. The state bureau of apprenticeship works together with other state agencies involved in human resource development, as well as the state's system of vocational-technical schools, to provide apprenticeship opportunities.

CONNECTICUT'S PREAPPRENTICE PROGRAM

The preapprentice training program allows school youth who are at least sixteen years old to start their apprentice training program on a part-time basis. The preapprentice registers with a program sponsor (employer) under the same conditions as an individual registering for a full-time apprenticeship. However, there is no provision for incremental wage increases, and a preapprentice can be paid a minimum wage. The preapprentice must also obtain parental consent to enter the program. Hours the preapprentice accumulates after school, on weekends, and in the summer are credited hours toward completion of an apprenticeship should they choose to pursue that particular field after graduation from high school. The duration of preapprenticeship must not exceed 2,000 hours, or twenty-four months, per regulation.

Firms that utilize apprentices in the machine tool and metal trades are eligible for a tax credit against their corporation business tax. Under Public Act Number 79-475, firms with machine tool and metal trades apprenticeship programs approved by the Connecticut Labor Department may receive a tax credit up to $4,800.00 for each eligible apprentice in a claim year. Apprentices can be utilized for tax credit purposes for the first three years in a four-year program.

We have also discussed a three-tiered approach to apprenticeship, which leads to promoting a continuity of training for our youth and helps to reach those youngsters oftentimes most needing the benefits of apprenticeship. Again, preapprenticeship opportunities are extended to the youth while still in secondary school. Time earned in training is then applied to a full apprenticeship, combined with higher education opportunities to earn a degree in a technical area in industry demand. Figure 6.3 describes the U.S. Navy's approach to this cooperative apprenticeship arrangement.

To fully appreciate and understand how an educational organization can facilitate initiation of an apprenticeship program, I am going to present a

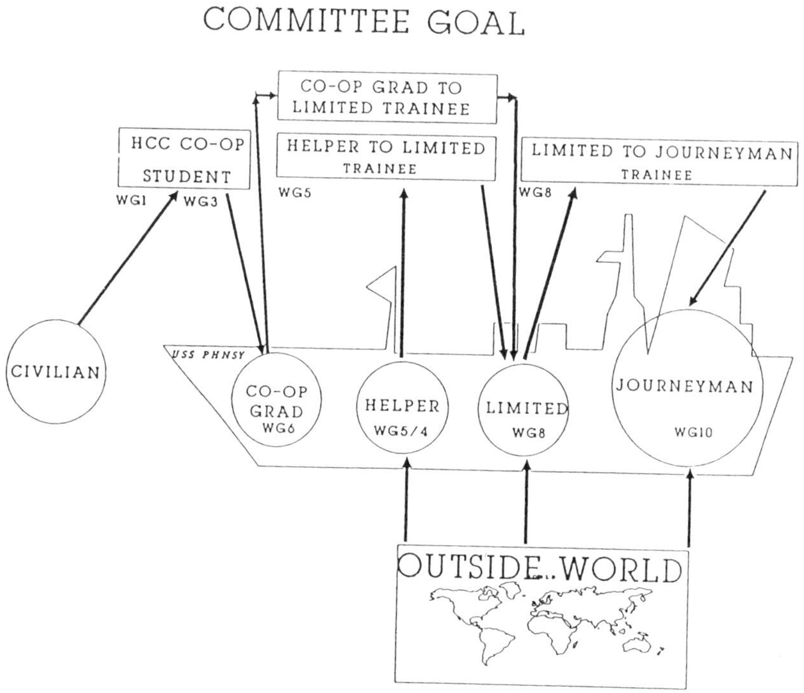

Figure 6.3 U.S. Navy cooperative apprenticeships.

scenario based upon some real experiences. The Danbury Area Community College (DACC) has been involved in providing services to employers and the state department of labor for apprenticeship for most of its thirty years of service to the community. This service has included the traditional kinds of activities: related classroom instruction; some referrals of students to employers for apprenticeship training; and occasionally, testing and screening of workers for apprenticeship when specifically requested by an employer. Additionally, the DACC serves as the area vocational-technical school for DACC's tri-county service area. The current president and board of trustees of Danbury Area Community College have established as primary missions and goals of the college:

(1) To provide more direct services to employers in work-based training
(2) To become proactively involved in local community economic development

Both of these goals are very admirable and, in fact, probably mirror most community colleges nationwide.

Danbury Area Community College convened its college advisory committee and requested its assistance in surveying area business and civic leaders to ascertain how it could begin to position the college's personnel, resources, and capabilities to meet these goals. From this survey emerged the following:

(1) A local labor union and contractors' association expressed a wish to join forces with the college in its apprenticeship education and training program. This union/contractor association uses a formal five-year apprenticeship program combined with an Associate degree program as the basic entry-level training for its journeymen.

(2) A national automotive manufacturer invited the college to apply for and consider adopting its corporate apprenticeship cooperative education program, which provides area automotive dealers with training Associate degree–level technicians. The college would need to work with the local dealers and corporate manufacturer in establishing an educational foundation to serve as the focal point for such training.

(3) Local machine tool shops expressed a need for trained machinists. They offered to participate in the training through apprenticeship, if such training were done in accordance with their national industry standards; however, these shops are small, and the college would need to shoulder the burden of coordinating such training.

(4) Area school systems expressed an interest in joining with the college to provide structured preapprenticeship school-to-work experiences for the youth of the community. The schools wish these experiences to provide the youth with credit towards an Associate degree, as well as towards full apprenticeships so that students can have multiple career opportunities and paths towards higher education.

With these expressions of interest in hand, the educational professionals began to understand that multiple needs and configurations do exist in most communities, and therefore, the educational institution needs to be able to structure apprenticeship program administration in such a way as to be able to meet the needs of multiple constituencies and organizations. Here is one such design for apprenticeship to meet these needs.

Danbury Area Community College will be able to work effectively with the labor union/contractors' association since the JAC already exists (again see Textbox 6.1). DACC suggested that a college representative attend JAC meetings so that the college is kept in the communications link regarding program operations. The JAC accepted this offer.

For the automotive manufacturer's program, however, the JAC needed to be assembled. The college decided to use the services of the state department of labor (DOL) bureau of apprenticeship and training. The college prepares the apprentice agreement and application for sponsor for each apprentice and work site identified as a service to the automotive community, and then processes them through the DOL (see Figure 6.4). This will be discussed as the chapter proceeds.

SELECTION OF APPRENTICEABLE OCCUPATIONS

Your choice of occupations for apprenticeship training will be initially based upon your needs assessment. Appendix B provides a listing of occupations that are traditionally apprenticeable and that are currently recognized by either the federal government and/or selected state departments of labor. In this case, the local employers expressed problems in attracting qualified workers. It appeared that the local schools were not producing students who had an interest in the trade or the prerequisite skills to master the trade through traditional on-the-job training.

While machine tool processes appear on the list, where a documented need exists for apprenticeship training and the specific occupation is not called out on the list in Appendix B, then the vocational-technical school and/or community college and firm in partnership will gather some preliminary information and contact the state's apprenticeship director or supervisor (Appendix C is a state DOL listing; Appendix D is a listing of U.S. Department of Labor state offices). This data includes:

- description of the specific jobs and duties performed in the occupation
- names and employer locations where this occupation is performed and where additional personnel will be a benefit
- estimated length of time needed to learn this occupation
- sources of related training for this occupation

The DOL representative will discuss the feasibility of indenturing apprentices in this occupation with the partnership.

FUNDING TO SUPPORT APPRENTICESHIP

Funding opportunities for program development is also a factor that should be discussed. A key decision maker in funding is the JAC. A training trust fund can be established, wherein the journeymen pay a tax that is used for apprenticeship training.

Some states have provided corporate tax credits to fund apprenticeship. For instance, in 1979, Connecticut began offering a corporate tax credit incentive to manufacturers for machine tool and metal trades apprenticeships. In 1995, this credit was expanded to the plastics industry. Presently, the maximum allowable credit is $4800 per registered apprentice (at least a

Not Valid Unless Approved by DILHR

WISCONSIN DEPARTMENT OF
INDUSTRY, LABOR AND HUMAN RELATIONS
JOBS, EMPLOYMENT AND TRAINING SERVICES DIVISION
Bureau of Apprenticeship Standards
Box 7972 Madison, Wisconsin 53707

Ed ____
Appr Dist ____
Trade Code ____
Committee ____
School Dist ____
Emp No ____
Dist Rep ____
FEIN ____

APPRENTICE INDENTURE

This Indenture prepared by ______________ Date ______________

THIS INDENTURE, Made in quadruplicate between ______________________
(Name of Employer)

hereafter called the first party, and ______________________
hereafter called the second party. (Name of Apprentice)

WITNESSETH, That the first party agrees to employ the second party as an apprentice
______________________ upon the terms and conditions in this indenture.
(Trade or Craft)

That the apprenticeship term begins on this date ______________________
(Month, Day, Year)

and terminates upon the completion by the apprentice of ______________________

____
(Term of Apprenticeship)

of employment for said employer in said trade, craft or business.

That the said apprentice agrees to diligently and faithfully fulfill all the obligations of this apprenticeship.

Social Security No. ______________

Date of Birth ______________

____
(Apprentice Legal Name - Print or Type)

____
(Street Address)

____
(City) (Zip Code)

____
(Firm or Corporation Name)

____
(Street Address)

____
(City) (Zip Code)

The provisions binding on the parties hereto are contained in exhibit "A" which said exhibit is made a part hereof.
The Department of Industry, Labor and Human Relations may annul this indenture upon application of either party after a satisfactory showing of good cause.
The Department of Industry, Labor and Human Relations shall issue a certificate of apprenticeship to the apprentice who has satisfactorily completed the terms of this indenture.
The apprentice's signature authorizes the school to release progress, grades and attendance reports to the Department and the signatory to this agreement while the agreement is active.

IN WITNESS WHEREOF, The parties have caused this indenture to be signed as required by Chapter 106.01 of the laws of Wisconsin.

X ______________________
(Apprentice Signature)

✓ ______________________
(Official Signature)

____
(Parent or Guardian Signature, if apprentice is a Minor)

____
(Official Name and Title - Print or Type)

JETA-4224 (R.05/92)

Figure 6.4 Sample indenture agreement (Wisconsin). (Reprinted with permission from the Wisconsin Dept. of Industry, Labor and Human Relations.)

NORTH CENTRAL HEALTH CARE FACILITIES
WAUSAU WI 54401

EXHIBIT "A"
MAINTENANCE MECHANIC
BUILDING & FACILITIES

EXTENT OF PERIOD OF APPRENTICESHIP: The term of apprenticeship shall be four years of not less than 8,000 hours. The first six months shall constitute the probationary period, but in no case shall this period extend beyond six months. Hours of labor shall be the same as established for skilled employees in this trade.

SCHOOL ATTENDANCE: The apprentice shall attend the appropriate technical college for four hours per week or the equivalent, for a minimum of 432 hours unless otherwise approved by the Bureau of Apprenticeship Standards. Hours of school attendance shall be counted as hours worked and the employer shall pay the apprentice the same rate of pay for these hours as hours on the job.

SCHEDULE OF PROCESSES TO BE WORKED: The apprentice, in order to obtain a well-rounded experience thereby qualifying the apprentice as skilled, shall have experience and training in the following, but not necessarily in this sequence nor all given at any one time.

PROCESS	APPROX. HOURS	CREDIT HOURS	BALANCE REQUIRED
Safety	300		
Lockout/tagout, single source			
Lockout/tagout, multiple source			
Eye protection			
Hearing protection			
Gloves			
Masks and respirators			
Ladders			
Scaffolding			
Power tools			
Hazardous chemical handling requirements			
General Building Maintenance	3368		
Lighting, lamps and fixtures			
Switches, receptacles, GFI receptacles			
Painting, caulking			
Wallcovering repairs			
Drywall repairs			
Ceilings			

Figure 6.4 (continued) Sample indenture agreement (Wisconsin). (Reprinted with permission from the Wisconsin Dept. of Industry, Labor and Human Relations.)

NORTH CENTRAL HEALTH CARE FACILITIES
MAINTENANCE MECHANIC BUILDING & FACILITIES
PAGE 2

	APPROX. HOURS	CREDIT HOURS	BALANCE REQUIRED
Masonry repairs Floor coverings Carpentry Furniture, cabinet repairs Roof maintenance Signs and pictures Air filter changing			
<u>Equipment Inspection & Adjustment</u> Check and adjust various belt drives Check and adjust chain drives Lubrication-various equipment Alignment, belt drives Alignment, chain drives Alignment, couplers Adjust valve and pump packings Kitchen and laundry equipment checks Carts and electric carts Wheelchairs and electric wheelchairs	300		
Equipment Installation and Repair Fans, blowers, air handling equipment Replace drive components Replace pump seals, couplers and other components Service and repair laundry pneumatic systems Power doors Elevator basics Air compressors	400		
<u>Piping</u> Toilets Sinks Showers and tubs Disposers Drain maintenance Grease traps Solenoid valves Domestic hot water systems	400		

Figure 6.4 (continued) Sample indenture agreement (Wisconsin). (Reprinted with permission from the Wisconsin Dept. of Industry, Labor and Human Relations.)

NORTH CENTRAL HEALTH CARE FACILITIES
MAINTENANCE MECHANIC BUILDING & FACILITIES
PAGE 3

<u>Boilers</u>	500
Basic boiler maintenance	
Chemical treatment systems	
Steam trap maintenance	
Steam, hydronic, airhandling	
Steamers, steam kettles	
Overall building environmental systems	
<u>Door and Cabinet Hardware</u>	150
Fire barrier hardware systems	
Closers and holders	
Security components	
Cabinet components	
Weather strips, thresholds, bumpers	
<u>Mandated Maintenance Requirements</u>	150
NFPA code requirements	
Hazard communication	
ILHR 10	
Other related topics	
<u>Refrigeration Maintenance</u>	100
Refrigerators	
Walk-in's and reach-in's	
Ice machines	
Chillers and cooling towers	
<u>Electrical Wiring</u>	600
Wire and cable runs (including phone, computer)	
Appliance hook-up	
Replacing ballasts and fixtures	
Electrical safety checks	
Motor Hookups - Single Phase	
3 Phase	
<u>Other Subjects</u>	
Blueprints and schematics	200
Maintenance record keeping	100
Emergency Power System	100
Cutting and welding	100
Miscellaneous trade related work	800
Related instruction	432
TOTAL	8000

The above schedule is to include all operations and such other work as is customary in the trade.

Figure 6.4 (continued) Sample indenture agreement (Wisconsin). (Reprinted with permission from the Wisconsin Dept. of Industry, Labor and Human Relations.)

NORTH CENTRAL HEALTH CARE FACILITIES
MAINTENANCE MECHANIC BUILDING & FACILITIES
PAGE 4

MINIMUM COMPENSATION TO BE PAID:

1st six months $7.03 per hour
2nd six months $7.35 per hour
3rd six months $7.72 per hour
4th six months $8.09 per hour
5th six months $8.39 per hour
6th six months $8.69 per hour
7th six months $9.08 per hour
8th six months $9.46 per hour

The present skilled rate is $9.81 per hour. If at any time the wage rate rises or falls, the apprentice's wagcc will be adjusted proportionately.

CREDIT PROVISIONS: The apprentice, granted credit at the start or during the term of the apprenticeship, shall be paid the wage rate of the pay period to which such credit advanced the apprentice.

School Record and Courses:

Work Record: (Work performed, employers, dates of employment)

NONE
Work Credit

NONE
School Credit

NONE
Total Credit Hours/Months

To be applied to the term of apprenticeship

SPECIAL PROVISIONS: The apprentice will complete the required number of hours for the Standard Red Cross First Aid and Personal Safety Certificate.

Figure 6.4 (continued) Sample indenture agreement (Wisconsin). (Reprinted with permission from the Wisconsin Dept. of Industry, Labor and Human Relations.)

two-year apprenticeship). The state's motivation for this tax credit was the rising age of the average craftsperson (fifty-years old) and little prospects of correcting the problem of a shrinking skilled workforce for manufacturers without such an incentive (Guerrera, 1995).

Other sources of potential funding include state vocational education programs, the Perkins Act, and the Youth Apprenticeship Act.

DEVELOPING AN APPRENTICESHIP DESIGN DOCUMENT

An apprenticeship design document is the formal charter of apprenticeship between all parties cooperatively sponsoring the apprenticeship program. This section of the chapter will discuss and describe the contents of such a document.

For the machine tool association, the college needed to become the catalyst for the formation of an educational foundation. It was decided that the organization to be developed take the form of a stand-alone educational foundation incorporated as a nonprofit corporation assembled for the sole purpose of apprenticeship training for its member firms and organizations. The educational organization would be a partner for apprenticeship training. To commence the development of the machine tool cooperative apprenticeship program, the initial step was to establish the foundation's board of directors, which would concomitantly serve as a JAC. It was to be equally constituted with member representation from employers and workers (or management and labor). [Where a union is involved, this is sometimes stipulated as the local union (labor or workers) and employer (or government agency).] I again stress that, to the degree possible, the school and/or college should be represented (even if *ex officio*) on the committee for coordination purposes. Local JACs must be endorsed by their state and/or national counterparts where such exist. For example, the IBEW New York City Local's JAC would be endorsed and recognized by the IBEW national JAC. In this case, the state apprenticeship council was asked to review the charter and apprenticeship design plan.

The corporation will have a board of directors and an administrator. The corporation will be set up under the provisions of the IRS Tax Code 501(c)(3)—a nonprofit educational foundation. The corporation will then establish the policies by which the program will be operated. The advantages of this form of organization were previously cited in Chapter 2.

The apprenticeship design document for the apprenticeship program underwritten by the educational foundation will now be highlighted and discussed in this chapter. The text that is set off and indented from the left margin is the design document text that should appear in your own design document.

The apprenticeship design document will include all of those components discussed and described in Chapter 5, which is the foundation for an apprenticeship program and which is highlighted in this chapter. It should open with a preamble that sets forth the purposes of the project, for example,

> *Preamble*
>
> The Danbury Area Community College–Business Alliance Educational Partnership, consisting of the Danbury Area Community College and the Danbury Area Machine Tool Industry Association, recognize their mutual responsibilities to provide training to support the future labor needs of the industry and the economic welfare of the Danbury area. Apprenticeship is a means to provide training for the occupations in the machine tool industry collectively in partnership with business and industry and the vocational-technical school and community college.
>
> The following standards of apprenticeship have been prepared by the joint apprenticeship committee (JAC) representing the Danbury Area Machine Tool Industry Association (DAMTIA) and the combined vocational-technical school and community college—The Danbury Area Community College (DACC). These standards are intended to provide a thorough training through work experience and related technical instruction for those individuals who wish to become skilled machinists in the industry.

EQUAL OPPORTUNITY PLEDGE

Every program of apprenticeship that will become registered with a state department of labor or the U.S. Department of Labor will contain a statement such as this:

> The recruitment, selection, employment, and training of apprentices during their apprenticeship shall be without discrimination because of race, color, religion, national origin, or sex. The sponsor will take affirmative action to provide equal opportunity in apprenticeship and will operate the apprenticeship program as required under Title 29 of the Code of Federal Regulations, Part 30.

ADMINISTRATIVE SUPERVISION

The elements included in the apprenticeship design document for the Danbury Area Community College–Business Alliance Educational Partnership will include the following:

> *Joint Apprenticeship Committee:*
>
> There is hereby established a joint apprenticeship committee. This committee shall be composed of members of the educational institution (DACC) and the industry association (DAMTIA). Operating as "The

Danbury Area Community College–Business Alliance Educational Partnership," hereinafter referred to as the "Corporation," it will be the employer of record for the apprentices.

Operating in this manner ensures that the participating employers will not need to bear the burden of overhead employment costs.

ADMINISTRATIVE PROCEDURES OF THE JOINT APPRENTICESHIP COMMITTEE

The first order of business for a JAC is then to select a chairman and secretary. In this organization, the following statement exists:

A. The executive director of the foundation will serve as the chair of the joint apprenticeship committee.

B. A quorum is required for a meeting; four (4) members of the JAC, including no fewer than two (2) workers and two (2) DAMTIA members, shall constitute a quorum.

Usually, there are six (6) members of a JAC in total. In this case, there must be at least two members of labor and two of management at all officially called meetings. An *ex officio* vocational-technical school and/or community college member is also recommended. The JAC should then make decisions about meeting times and places.

Duties of the Joint Apprenticeship Committee:

A. To meet monthly to discuss and make recommendations in connection with the training of apprentices

B. To develop and refine a curriculum of study and job experience for apprentices in each discipline

Adjustment of Differences:

In the event that differences arise that cannot be satisfactorily settled by the parties to the agreement, either party has the right and the privilege of referring to the state apprenticeship council for consultation.

A. To see that the proper records are kept of the apprentices' progress on the job, as well as in related classroom instruction

B. To cooperate with the state apprenticeship council in awarding certificates of completion of apprenticeship

In this program, the foundation will grant a certificate of completion that will indicate that the apprentice has met NTMA industry standards. This aspect of the program completion ensures work portability for the journeyman. Additionally, the college will grant the Associate degree upon degree requirements completion.

C. To provide overall administrative guidance to the apprenticeship program

One of the neat features of the apprenticeship is its ability to change so as to continually keep up-to-date. This statement makes this both possible and desirable:

Modification of Standards:

These standards may be amended or modified at any time by the committee whenever such change is considered beneficial to the apprentices' training program, provided, however, that such changes will not affect apprentice agreements then in effect, except upon the mutual consent of both parties to the agreement. All such changes or modifications will be submitted to the state apprenticeship council for approval and registration. These standards are considered to be part of this agreement with the same force and effect as if specifically included herein.

RULES AND REGULATIONS OF APPRENTICESHIP

The apprentice signs an apprenticeship indenture (Figures 6.4 and 6.5), which is a written agreement between apprentice, the employer, and the state (e.g., state of Connecticut, division of apprenticeship). The agreement states that the apprentice will work for the employer for a specified period of time and that the employer will pay a specified wage and provide the apprentice with specified skills and competencies.

Individual Apprentice Agreements:

Each apprentice, together with the foundation, shall sign an apprenticeship agreement in triplicate—one copy to be registered with the state apprenticeship council, one copy for the Foundation, and one copy for the apprentice.

Official Approval and Registration:

These standards shall be approved by formal action of the JAC and shall be approved and registered with the state apprenticeship council.

RECRUITMENT, SCREENING, AND SELECTION OF APPRENTICES

Again, the qualifications required for a student to enter into an apprenticeship program must be clear and objective. Fairness to all students must be

AT-5 (Rev. 05/95)

RVTS CODE

CONNECTICUT STATE APPRENTICESHIP COUNCIL

Connecticut Department of Labor

Apprenticeship Agreement

This agreement entered into this _________ day of ______________________________________,19________

between ____ hereinafter referred to as the Trainer
(Name of program sponsor)

and ____hereinafter referred to as the apprentice for
(Name of Apprentice --typed or printed)

the purpose of learning the skills in the Trade of ____

Full term of Apprenticeship ____________ Credit for prior trade experience ______________________
(Hours) (Months) (R.I.) (OJT) (Months)

Graduate of ____ High /Technical School in __________
(Year)

In conformity with the program sponsor's Standards as approved or amended in accordance with the Commissioner of Labor's Work Training Standards for Apprenticeship and Training Programs are as follows:

The Trainer agrees to employ the Apprentice for the purpose of enabling said Apprentice to learn and acquire the trade or craft upon the terms and conditions contained in the standards without discrimination because of race, color, religion, national origin, age, physical disability or sex.

The Apprentice agrees to perform diligently and faithfully the work of the trade or craft complying with the training program contained in the standards and to attend required related instruction.

Graduated scale of wages to be paid the apprentice. Period (Months,Hours,Years)________________
(No.) (Period)

1st __________%	5th __________%	9th __________%
2nd __________%	6th __________%	10th __________%
3rd __________%	7th __________%	11th __________%
4th __________%	8th __________%	12th __________%

Current minimum completion rate: $____________ per hour

NOTE: Pursuant to U.S.C.F.R. Title 29, Part 5 and C.G.S. 31-53, apprentices assigned to Prevailing Wage Project jobs must be paid their percentage of the program sponsor minimum completion rate or the project journeyperson's rate (Prevailing Wage), whichever is higher, plus 100% of the fringe benefits listed in the wage determination for their occupational classification.

☐Male ☐Female Apprentice Social Security #______________________________ Birth Date ____________

Highest grade completed __________ ☐Veteran ☐Non Veteran

☐White ☐Black ☐Asian or Pacific Islander ☐Native American ☐Hispanic

Apprentice (Signature)	Program Sponsor
Address	Address and City
City	Representative of the above (Signature & Title)
Parent or Guardian (Signature) (for minor age 16 and 17)	Approved by the Connecticut State Apprenticeship Council Chairperson, Deputy Labor Commissioner

Figure 6.5 Sample indenture agreement (Connecticut). (Reprinted with permission from the Connecticut Dept. of Labor.)

demonstrated. The preapprenticeship component of this program will work to guarantee this. By working with the local secondary schools to provide preapprenticeship opportunities for the community's youth, all will have an opportunity at the full apprenticeship. The statement in the design document states:

> *Apprentice's Entrance Qualifications:*
>
> Admission requirements should be printed and published and distributed to all who request it. This apprenticeship program has a probationary period of six months, during which time the foundation reserves the right to dismiss apprentices who do not demonstrate reasonable work habits and technical progress.
>
> A. Applicants must have reached their eighteenth birthday before the expected start of the full apprenticeship. However, applicants for the preapprenticeship must have reached the age of sixteen, and have the written permission of a parent or guardian.
>
> B. All applicants must be citizens of the United States.
>
> C. Minimum entrance requirements:
>
> (1) High school diploma (or satisfactory progress towards same with a GPA of 2.5 for the preapprenticeship)
>
> (2) High school course requirements include:
> - one course in drafting or mechanical drawing
> - two years of algebra
> - one year of geometry
> - four years of English
> - one year of physics
>
> D. All applicants must submit a high school transcript signed and officially sealed by the school official, together with an application for apprenticeship.
>
> E. Selection of apprentices shall be on the basis of qualifications alone, without regard to race, creed, color, sex, or national origin. The operation of the apprentice program will be on a completely nondiscriminatory basis.
>
> F. All applicants must pass a physical examination given by a physician of the foundation's choosing.
>
> G. To be accepted as apprentices, applicants will be interviewed by a committee of foundation representatives.
>
> H. Related classroom training will be scheduled by the JAC and college and will be after working hours. The apprentice is required to attend

classes as a condition of employment. Satisfactory progress in classwork is required to maintain the apprenticeship.

I. All apprentices will conform to foundation rules and regulations.

DESIGN ON-THE-JOB AND RELATED TRAINING

In this program it is decided that:

- Apprentices will receive a minimum of 640 hours of shop training prior to placement on the job. This training will be done through the vocational-technical school.
- Apprentices will be rotated through employer shops to ensure that each apprentice is able to be exposed to and master all of the work processes associated with the trade. The administrator will determine those shops to be used for each work process phase to ensure that those shops best able to provide specific experiences are used efficiently. Shopowners will need to agree to these terms if they are to be part of the program.
- The related education and training portion of the program will be delivered by the participating community college—herein Danbury Area Community College in a bona fide Associate degree program.

In this program, periodic review and counseling of students will be done after each rotation.

On-the-Job Training:

Instruction and experience will be given in the major work process with the understanding that rotations and the flow of work during apprenticeship will determine the order in which these work processes are undertaken. It is further understood that, in the event a major work process or a phase thereof becomes unavailable due to the nature of work at hand during the apprenticeship, such omission will not affect successful completion of the apprenticeship.

Related Classroom Instruction:

Apprentices are required to attend related classroom instruction during the term of apprenticeship. Hours spent in class will not be considered work experience but will apply toward completion of the program.

Again, the information to be imparted in the classroom should emanate from the task analysis supporting the occupation in question. The design document statement reads: “The courses to be taken at Danbury Area

Community College are specified [and based upon the Community College of Rhode Island's program for MechTech, Inc., of Rhode Island]. These courses are to satisfy the related training portion of the program standard."

Machine Processes
A.A.S. Degree Program

Course Title	Credits
Manufacturing Technology	9
Diemaking I & II	6
CNC Programming	2
Machinery Handbook	2
Blueprint Reading	2
Technical Drawing	2
Liberal Arts Electives	9
Machine Tool Geometry	2
Algebra	3
Strength & Properties of Materials	2
Technical Physics	4
Cost Estimating	3
English Composition	3

Program Display:

Attached hereto is the apprentice program display of work processes and schedule for the occupation selected (see Figure 6.6).

Term of Apprenticeship:

The term of apprenticeship will be 8,000 working hours and the successful completion of all academics that are required as part of the Associate degree.

The term of apprenticeship is determined by summing the time it takes to master each work process plus related instruction (see Chapter 5). The statement in the design document states: "This program will take approximately 8,000 hours of on-the-job instruction (four years) plus a minimum of 144 hours a year of related instruction (576 hours) to complete." A full-time apprenticeship (fifty-two weeks) is 2,080 hours; therefore, the program will be four and a quarter years long. Typically, for a program such as this, the related instruction will be factored into the hours required for mastery of the process. What could prolong the term of apprenticeship,

A. Bending Floor Mechanic	**2,470**
1. Heating Torch Training and Operation	20
2. Straighten Foundations and Assembled Units	1,000
3. Line Heating Shell Plates	520
4. Training and Operation of Furnaces	100
5. Heat Treating Processes Applied by Bending Floor	200
6. Twisting of Plates and Shapes From Furnace and Line Heat	230
7. Shape Various Materials with Sets and Bending Floor Forming Equipment	200
8. Development and Construction of Bending Floor Jigs and Fixtures	200
B. Machine Operations	**2,470**
1. 500 Ton Press Forming and Straightening	280
2. 300 Ton Press Forming and Straightening	240
3. 50 Ton Press Forming and Straightening	150
4. 75 Ton Horiz. Press Forming/Straightening	150
5. Radial Drill Press	200
6. Horizontal Do-All Saw	200
7. Vertical Band Saws	50
8. 32' Bending Rolls - Forming Shell Plates and Various Shapes	250
9. Niagara Punch	40
10. Beatty and Kling Punch	80
11. Piranha Punch	80
12. Beveling Machine	80
13. 16' - 400 Ton Press Brake	250
14. Bulldozer - Horizontal Bender	50
15. Horizontal Frame Bender	250
16. 14' and 4' Gate Shears	50
17. Door Rolling Machine and Trimming Shears	40

Figure 6.6 Schedule of work processes (Bath). (Reprinted with permission of Bath Iron Works Corporation.)

C.	**Fabrication Hardings**	400

1. Layout — 120
   a. Helping Journeymen layout plates from templates
   b. Helping Journeymen layout shapes from templates, sketches and marking symbols
2. Small Assembly — 280
   a. Helping Journeymen assemblers layout from blueprints, check material and the tack welding of small assemblies peculiar to the Hardings assembly floor. This includes aluminum and steel foundations, webs and structures.

D. Hull Straightening - Bath (Dept. 39) — 160

1. Helping Journeymen straighteners on decks, bulkheads, shell and other related straightening functions of the Bath operation.

E. Related Classroom Instruction — 500

1. Apprentice will receive approximately four hours per week of related classroom instruction throughout the term of apprenticeship.
   a. Instruction in classroom including Mechanical Drawing, Development, Mathematics, Strength of Materials, Mechanics and other shop and educational courses.
   b. Extension classwork, including practical demonstrations in shops and on ships which cannot be given in the classroom.

Total Hours 6,000

Figure 6.6 (continued) Schedule of work processes (Bath). (Reprinted with permission of Bath Iron Works Corporation.)

however, would be the courses required by the Voc-Tech or community college for completion of a program, coupled with a degree or certificate, as will be discussed shortly. Another typical statement of the apprenticeship term might be

> The bricklayer apprenticeship program is three years in length. Four hundred hours of related instruction is required. This is based on a four hour work experience per week, or the equivalent, and is paid for by the employer. The apprentice will attend evening school and take such courses as the committee requires. A total of 110 hours of evening school are required. This will be on the apprentice's own time and at his own expense. Some of the courses required are as follows: first aid and personal safety, beginning stick electrode arc welding, OSHA, builders level transit. (Wisconsin Bricklayer)
>
> *Field of Specialization:*
>
> Apprentices will be appointed to the apprentice program in a field of specialization as mutually agreed upon between trainee and the foundation.
>
> *Credit for Previous Experience:*
>
> At the discretion of the JAC, up to 1,000 hours of credit may be given to successful applicants for apprenticeship. In the case of preapprenticeship, this is by prior agreement. Any credit will be applied toward the work process portion of apprenticeship. If such credit is granted, alternative work experience shall be provided as a substitute at the discretion of the JAC. There will be no academic credit awarded.
>
> *Hours of Work:*
>
> The hours of work for apprentices shall be the same hours as for other workers at the employer's shop for which the apprentice is serving a rotation.

PLAN FOR APPRENTICE EVALUATION

A process for credentialing the training process and the worker's training accomplishments is to be devised. Competency-based examinations will be used for this purpose. Chapter 9 discusses how to do this.

As more and more community colleges begin to underwrite and guarantee a learner's competence exiting from such training, program evaluations taken from your training standards will form a backbone for and will serve as a foundation for evaluation. The standards will also provide a framework for how this program will be reviewed periodically and how apprentices will be monitored and evaluated in both job performance and related instruction. Many apprenticeship programs tie evaluation into the function of the local JAC. Standards must include how apprentice records and progress reports will be kept.

Note that a major issue is confidentiality of records. The U.S. Privacy of Information Act prohibits releasing information relating to apprentice progress to anyone outside of the educational institution (and inside as well). Therefore, if your college is a training agency for your apprentice program, a release-of-information form will need to be devised and administered so that course grades and related information can be given to the JAC and firm as needed. This might not include college credit grades if these courses are outside the scope of the apprentice program.

> *Probationary Period:*
>
> A probationary period will be served by all apprentices. This probation will be in effect from the beginning of academic classes until completion of fourteen weeks of class. During this period, the apprentice agreement may be cancelled by either party thereto by notification, in writing, of such a desire. Notice will be given to the state apprenticeship council.
>
> Work experience will be evaluated by the employer's supervisor at the end of every rotation or eight weeks, whichever comes first. Evaluation will carry a letter grade. A grade less than C will be considered failure in a work process. Hours worked in a failed grading period will not be credited towards program completion. The apprentice must repeat the rotation in order to earn credit. Failure in the second experience will result in termination from the program.
>
> Academic performance will also be evaluated according to the DACC rules and regulations. Failure to maintain a C academic average after a prescribed DACC academic probation period will result in termination from the program.

Progressive wage increases should be tied to successful evaluations.

> *Apprentice Wages:*
>
> The pay rate for an apprentice shall be in accordance with the schedule specified in Table 6.2.

For programs that are cosponsored with a labor or employee organization, the progressive wage schedule for the apprentice should be negotiated between labor and management. This is the case with the educational foundation that will serve as the employer of record. A progressive wage schedule was developed based upon apprentice hours worked.

TABLE 6.2. **Progressive Wage Schedule.**

Wage Structures									
Apprentice Work Hours	Hourly Wage	Hol.	Vac.	Taxes	Work. Comp.	Med.	Educ.	Ovhd.	Company Charge
Preapprentice Co-op Program									
H.S. Jr. Summer	$6.00	X	X	.25	.25	X	X	.75	$7.25
H.S. Sr. Fall	$6.50	X	X	.26	.26	X	X	.75	$7.77
H.S. Sr. Winter	$7.00	X	X	.32	.32	X	.5	.75	$8.89
Registered Apprentices									
0–1,000	$7.25	.10	X	.52	.32	1.00	.5	.75	$10.44
1,001–2,000	$7.50	.12	.15	.54	.32	1.00	.5	.75	$10.88
2,001–3,000	$8.00	.13	.17	.57	.36	1.00	.5	.75	$11.48
3,001–4,000	$8.50	.14	.19	.58	.39	1.00	.5	.75	$12.05
4,001–5,000	$9.00	.15	.20	.60	.42	1.00	.5	.75	$12.62
5,001–6,000	$9.50	.16	.23	.62	.45	1.00	.5	.75	$13.21
6,001–7,000	$10.00	.17	.25	.65	.47	1.00	.5	.75	$13.79
7,001–8,000	$10.50	.18	.30	.68	.49	1.00	.5	.75	$14.40

NUMBERS OF APPRENTICES

The numbers of apprentices that should be taken on for training is an important aspect of apprenticeship program development. You should discuss this with the firms' principals. A business should be looking at its existing workforce, as well as its workforce needs four to ten years away. Apprenticeship permits the firm to train future workers in the skills that its present workers have. It permits development of a "farm team" for future use. This aspect of program development is stressed by the automotive manufacturers in their apprenticeship program models. Union programs also express this as a major concern. The design document states: "To do this the initial steps to follow are to 1) define your present supply of skilled workers, 2) project the numbers of workers you will need for the future, and 3) project the numbers that will be trained under the present program." Your standards should also specify the ratio of apprentices to journeymen that is consistent with your needs and continuity of labor force.

> *Numbers of Apprentices:*
>
> The foundation shall establish a number for apprentice indenture that corresponds to no more than two per employer electing to sign an agreement with the foundation to participate.

DESIGN A CERTIFICATE OF COMPLETION

Completion of all the requirements of the registered apprenticeship program entitles a graduate to receive a state certificate of completion and often is also recognized for college credit towards an Associate degree. This certificate attests to the attainment of certain minimum standards and is a passport to jobs all over the country. The craftsperson who successfully completes both the work processes and related instruction required in the cooperative apprenticeship acquires status in the trade with others who have had similar experiences and completes a significant portion of a college degree, a necessary prerequisite for assuming an industry leadership role.

PLAN AND DESIGN FACULTY DEVELOPMENT

In apprenticeship, employers provide the jobs—which are referred to as the learning stations or work sites. The firm's senior skilled employees, called journeymen, are typically the on-the-job instructors. Therefore, Danbury Area Community College faculty will provide opportunities for the journeymen who are selected as instructors to develop needed adjunct faculty skills. Some states provide funding for in-service faculty development for their instructors. Certain professions also provide training assistance. For instance, in the firefighter professions, instructor development is recognized as essential. California has a series of instructor development courses that cover the basics of instructor training, leading to certification as a firefighter instructor. These courses also enable these individuals to be certified as community college instructors. Opportunities for staff development can be funded through Perkins and other funding mechanisms through a state department of labor. More about this is given in Chapter 8.

REVIEW APPRENTICESHIP AGREEMENT

At this point, the apprenticeship agreement should be reviewed by all concerned parties—employer, union, apprentice, and educational agency (if applicable). Ensure that all parties understand the commitments made and are prepared to fulfill their individual roles and responsibilities.

REFERENCES

American Association of Community and Junior Colleges (AACJC). (1991). "Colleges Moving to Guarantee Their Graduates," *Community Technical and Junior College Times,* III(22):1.

California Firefighter Joint Apprenticeship Committee. (1989). Descriptive Brochure. Sacramento, CA.

Cantor, J. A. (1992). *Delivering Instruction to Adult Learners.* Toronto, Canada: Wall & Emerson.

Cantor, J. A. (1990, April). "How to Perform a Comprehensive Course Evaluation," *Performance and Instruction,* 8–15.

Cantor, J. A. (1986). "A Systems Approach to Instructional Development in Technical Education," *Journal of Studies in Technical Careers,* IX(2):155–166.

Cantor, J. A. (1985a). "JTPA Supports the Shipbuilding Industry," *Journal of Studies in Technical Careers,* VII(3):205–213.

Cantor, J. A. (1985b). "Task Evaluation: Comparing Existing Curricula to Task Analysis Results," *Journal of Educational Technology Systems,* 14(2):157–163.

Electrical Joint Apprenticeship and Training Committee. (1989). *Local Apprenticeship and Training Standards for the Electrical Contracting Industry.* New York: Author.

Empire State College. (Undated). *Apprentice Handbook.* New York City, NY: Author.

Glover, R. W. (1986). *Apprenticeship Lessons from Abroad. Information Series No. 305.* Columbus, OH: The National Center for Research in Vocational Education, Ohio State University.

Grabowski, D. J. (1989). Testimony of Donald J. Grabowski, President, National Association of State & Territorial Apprenticeship Directors before the Commission on Workforce Quality & Labor Market Efficiency. National Association of State and Territorial Apprenticeship Directors, Northlake Community College, Irving, TX, May.

Guerrera, J. (1995, November). "Corporate Tax Credit for Machine Tool and Metal Trade Apprentices in Connecticut." *USAA Sentinel*. p. 1.

Martin, S. T. (1981). *Apprenticeship—School Linkage Implementation Manual*. Washington, D.C.: CSR, Inc.

MechTech, Inc. (1995). Program Materials and Personal Interviews. Cranston, RI.

New York State Department of Labor, Apprenticeship and Training Council. (1987). *An Overview of Apprentice Training*. Albany, NY: Author.

State of Connecticut, Department of Labor, Apprenticeship and Training Division. (1995). Descriptive materials and personal interview notes on Connecticut apprenticeship.

State of Wisconsin, Bureau of Apprenticeship Standards. (1991). *A Guide to Apprenticeship in Wisconsin*. Madison, WI: Author.

Tuholski, R. J. (1982, October). "Today's Apprentices—Tomorrow's Leaders," *VocED*, 37–38.

U.S. Department of Labor. (1989). *Setting Up an Apprenticeship Program*. Washington, D.C.: Bureau of Apprenticeship and Training.

CHAPTER 7

Recruiting, Screening, and Selecting Apprentices and Employers

THIS chapter is about identifying, selecting, and matching students and employers for apprenticeships. Employers and educators working in partnership through apprenticeship are in a unique position to bridge students and the workplace. This chapter will discuss how these partners can work together to identify, screen, test, counsel, and place students on the job. Information will also be provided about recruiting employers, screening work sites, and preparing work-site supervisors for the all-important roles as mentors to apprentices.

At a recent meeting of machine tool industry employers, representatives of the federal and state bureaus of apprenticeship, college educators, and economic development specialists, discussions ensued about how to form partnerships to identify, recruit, and train workers. Many pertinent issues surfaced. These issues included concerns about a lack of trained workers for this industry, problems in finding young people interested in being trained for the industry, employer interest in joining forces with education to identify and train young people for the workforce, funding to support training, and perceptions of parents and youth themselves of the trades as viable careers. Indeed, one might very well become discouraged after hearing all of these seemingly insurmountable problems.

But don't despair, for where there exist mutual benefits for those involved, these problems can be solved. First, as educational partners, you have the power to bring to the surface the mutual benefits to be derived by all parties concerned. Depending on the role that an educational institution presently plays in apprenticeship, involvement in trainee recruitment and employer recruitment will vary. Those providing related training support for apprenticeship are already involved in marketing these services to employers in the community—and thus in the mode of identifying new employers to benefit from these services. Most all employers will be happy to learn that an educational institution can be useful in providing the following kinds of services.

- apprenticeship promotion
- trainee recruitment
- trainee screening and testing
- trainee placement and counseling
- trainee evaluation on-the-job and in the classroom
- employer recruitment
- employer and work-site screening
- employer preparation

Each of these topics will be covered in this chapter.

PROMOTING APPRENTICESHIP

Have we as an American society become snobbish when it comes to counseling our youth about going into careers in the crafts and trades? The answer is probably yes. We do tend to suggest to the more academically successful student a college-bound course of study. To the student not making the grade, we promote a vocational course of study. The fallacy herein is that the technical occupations now require a very rigorous secondary school preparation, and these technical occupations are in high demand and command a very handsome salary.

Alternatively, those students relegated to the vocational courses of study oftentimes lack basic skills preparation, the motivation to succeed in the field of endeavor, and all too often a less than challenging curriculum in isolation of the real world. Businesspeople are not out of line when they argue that the students who come to them from the vocational school are less than ready to meet the demands of the workplace.

One alternative to this dilemma is cooperative apprenticeship. As has already been argued, through apprenticeship, employers and educators team up to provide quality workplace training, coupled with classroom education. However, cooperative apprenticeship demands the same high level of preparation as does a college education, and the student must be highly motivated to meet the challenge of an apprenticeship—which oftentimes takes as long to complete as a baccalaureate degree.

The need at this point is for society—parents, clergy, community leaders, and so on—to recognize in word and deed the value of highly technically trained men and women. A certificate of apprenticeship completion, coupled with a college diploma, must signify what it rightfully is—a dual accomplishment of excellence.

TRAINEE RECRUITMENT

Where should we look for qualified people who have both the interest and ability to pursue training and education through cooperative apprenticeship? One answer should be in our high schools. Through promoting preapprenticeship, we will increase the supply of properly prepared apprenticeship candidates. What can be done?

- Businesspeople and community-technical college educators must establish a working relationship with secondary educators who share an interest in promoting educational opportunities. Preapprenticeships can be developed that carry credit applicability for on-the-job training towards high school diplomas, Associate degrees, and full apprenticeships. Career awareness activities also help to point students toward the skilled trades and technical professions. These experiential learning opportunities for high school–age youth will provide exposure to the various technical career opportunities. What can be done?
- Community-based programs such as housing rehabilitation can provide real-world exposure for potential apprenticeship candidates. This is a good opportunity to establish an educational partnership. Secondary school students should have an opportunity to do volunteer work and/or gain part-time employment in the community in activities or experiences that expose them to career options and that can get them excited about apprenticeships.

What are good sources of adult-age trainees?

- Vocational-technical schools and community colleges are good sources for recruitment of adult-age students who are looking for career changes or students who are displaced from careers no longer in demand or homemakers looking to enter the workforce and/or separated military personnel seeking new careers.
- Other traditional sources for trainee recruitment include the state department of labor or other state human resource agency. The state department of labor will identify and screen potential applicants for apprenticeship. The bureau of apprenticeship representative can match applicants to employers who indicate an interest in sponsoring apprentices. The employment offices of labor unions, trade associations, large manufacturers or employers, and so on can do likewise. A vocational-technical school and/or community college can work out referral arrangements with these agencies for channeling of potential apprentices to the school or college for testing, screening and counseling for eventual apprenticeship placement. Figure 7.1 is an example of a brochure developed jointly by a state department of labor and a community-technical college system for promoting cooperative apprenticeship.

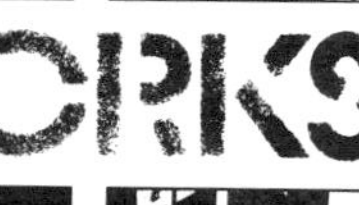

Figure 7.1 Department of Labor/technical college apprenticeship promotional brochure. (Reprinted with permission from the Wisconsin Dept. of Industry, Labor and Human Relations.)

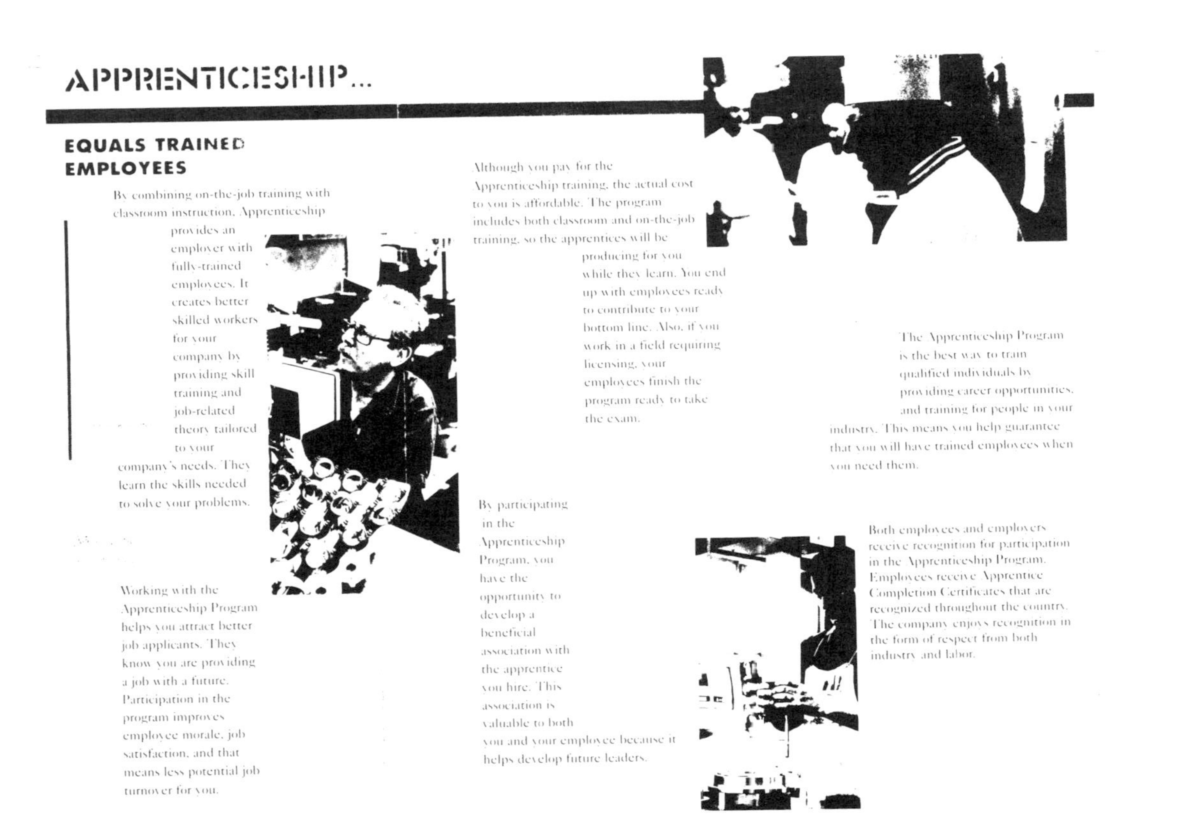

Figure 7.1 (continued) Department of Labor/technical college apprenticeship promotional brochure. (Reprinted with permission from the Wisconsin Dept. of Industry, Labor and Human Relations.)

ADMINISTERING A RECRUITMENT PROGRAM

Vocational-technical schools and/or community colleges can assist employers with the actual administrative details of apprentice recruitment in several ways. From developing and advertising the apprenticeship opportunities to the community at large to processing and screening of applications, interviewing, pretesting, and referring of applicants to employers, an educational institution can provide invaluable assistance to the business in the overall partnership.

Advertising Apprenticeship Opportunities

In order to ensure that apprenticeships are, in fact, extended to all people within the community without regard to race, creed, ethnicity, religion, sex or sexual orientation, or any other differences, apprenticeship opportunities should be announced publicly and a period of collection of applications should be extended. Figure 7.2 is an announcement that appeared in *The Washington Post.* From such an announcement, a large number of calls will undoubtedly be received; therefore, a system of dispensing and recording applications needs to be established by the educational partnership, and a timeframe needs to be established for receiving applications.

Designing an Application

There should be a standard application form to ensure consistency of applicant evaluation. The application package should include all of the information that the applicant needs to know about the application process for acceptance into the apprenticeship. This should include:

- applicant minimum personal qualifications (age, physical requirements)
- educational requirements, including courses needed to have been completed in high school
- required personal references (usually letters from two to three people such as teachers, clergy, other people in the community)
- required documents needed with application (e.g., high school transcripts, birth certificate, immigration visa)
- written essays used in order to test an applicant's written communications skills (Common essay subjects might include reasons for applying to apprentice school or reasons why you want to be a technician.)

Maintenance

CARPENTER APPRENTICE

Salary to $27,100

NAVY FEDERAL CREDIT UNION, headquartered in Vienna, VA has a position open in their Building Maintenance Shop for a Carpenter Apprentice to provide assistance in constructing, altering, installing, repairing and maintaining office buildings, equipment, and furnishings in the areas of carpentry to include walls, doors, windows, ceilings, tile, carpet, and specialty items such as cabinets and shelves. Qualifications are:

- Three years carpentry experience
- Successful completion of third year classes in the Apprenticeship Program for carpenters or formal training equivalent
- Knowledge of tools, mathematics, materials, and safety regulations associated with the carpentry trade
- Ability to follow work instructions and read and interpret simple drawings and diagrams
- Must possess a valid driver's license

We offer a stable work environment with excellent benefits to include health and dental insurance, free on-site exercise facility, free parking and opportunities for advancement. To be considered, call (703) 255-8800 to receive an application or send your resume to:

P.O. Box 3400
Merrifield, VA 22119-3400

We are committed to a drug-free workplace

EOE M/F/V/D Fax (703) 206-4510

Figure 7.2 Announcement—"call for apprentice." (©1995, *The Washington Post.* Reprinted with permission.)

- requirement for a personal interview if selected for further consideration

Figure 7.3 is a standard application form used by one apprentice school. Figure 7.4 is a standard form used for references.

SCREENING AND TESTING

Once applications are received and the cut-off date for submitting applications has arrived, it is time to review those received and separate out those that meet the minimum requirements for admission. If more meet this criteria than can be accommodated in a given period, then decisions about the relative merits of each qualified applicant now need to be considered and the very best candidates singled out and rank-ordered for consideration. These applicants will now be interviewed and tested.

INTERVIEWING CANDIDATES

Those applicants who emerge as best-qualified, based upon a paper review of applications, should be listed formally on a call for interviews report. In this manner, you will have a formal record of proceedings leading to the interview stage of the recruitment process. The full admissions committee (usually a subset of the JAC plus the educational institutions) will be present at all interviews. This is in the interest of fairness and consistency. Individual appointments of equal time length will be scheduled for each candidate. A list of questions to be posed to each and every candidate should be drafted by the committee. Usually, each committee member is charged with posing one or more questions to the candidate. Areas to question include:

(1) Why has the candidate chosen this apprenticeship program?
(2) Why does the candidate wish to learn this particular trade?
(3) What are the candidate's career aspirations?
(4) What have been the candidate's greatest accomplishments?
(5) What work experience has the candidate had in or related to this field?

The committee should look for indications of the candidate's verbal or speaking communications skills, presence, ability to relate to people, and so on. However, no questions of a personal nature or of a nature that suggests bias of any kind should be asked. The responses should be recorded either by tape and/or in writing for later review. A formal response checklist should be completed for each candidate by each committee member as a means of rating each candidate's interview for admissions purposes.

APPLICATION TO THE APPRENTICE SCHOOL

NN 1331 (REV 14) 02 09 93

NEWPORT NEWS SHIPBUILDING
A TENNECO COMPANY

4101 Washington Avenue
Newport News, Virginia 23607
(804) 380-3809

A $25.00 non-refundable Application Fee must accompany the application.

GENERAL INFORMATION

The Apprentice School is an operating department of Newport News Shipbuilding and Dry Dock Company. It is the policy of the school and company to provide employment, training, compensation, promotion and other conditions of employment based on bona fide occupational requirements and without regard to race, color, religion, national origin, sex, age, or disability.

ADMISSION REQUIREMENTS

To be considered for admission to the Apprentice School, an applicant must:

1. Have a high school education with a minimum of four units passed in any combination of the following: one or two units of algebra; one unit of geometry; one unit of chemistry; one unit of physics or principles of technology; one unit of mechanical drawing; one unit of vocational/ technical or computer science; or one unit of advanced mathematics. A unit is considered to be one year of high school study or its equivalent.

2. Be physically able to perform the essential duties required in the trade requested or assigned, as determined by the company's Medical Department.

3. Have a background which will allow the individual to be issued a government furnished security clearance.

4. Be at least 18 years of age and must not have reached his/her 30th birthday. High school seniors may apply before reaching age 18.

Figure 7.3 Newport News Shipbuilding "application to apprentice school." (Reprinted with permission from The Apprentice School, Newport News Shipbuilding.)

ADMISSION PROCESS

1. Application forms will be furnished by the Admissions Office upon request.

2. Applications are accepted daily, and selections are made on a monthly basis as openings occur.

3. To apply for admission, submit a completed application form along with the $25.00 (non-refundable) application fee to the Admissions Office. Checks for the application fee should be made payable to The Apprentice School.

4. Have at least two of your personal references submit written recommendations directly to the admissions office. Please have each reference include a telephone number at which he/she can be reached (Please use the reference forms and return envelopes provided).

5. Official high school transcripts and official transcripts from any post-secondary institutions you have attended must be forwarded from that institution directly to the Admissions Office.

6. Candidates selected to continue the admission process will be scheduled for a tour and personal interview and will undergo agility and/or strength or other testing and will be tested for alcohol and controlled substances.

7. Monthly selection meetings are held at which all qualified applicants are considered for available openings. Offers are extended to applicants which the Admissions Office determines to be best qualified for the openings available. Admissions are competitive, with more qualified persons applying than the number of openings available. Meeting the minimum requirements specified above, therefore, does not assure selection.

8. Potential apprentices will be scheduled for a physical examination administered by the company's Medical Department. Employment will be contingent on the results of the physical examination.

9. Potential apprentices will be scheduled for a security clearance interview administered by the company's Security Office. Employment is subject to the security policies of the company.

COMPLETE THE FOLLOWING:

"REASONS FOR APPLYING"

Write a brief personal statement covering the following topics:

a. What do you know about the Apprentice School?
b. What was the source of your information?
c. Why do you want to be admitted to the Apprentice School?
d. What are your plans and goals beyond completion of an apprenticeship?

Figure 7.3 (continued) Newport News Shipbuilding "application to apprentice school." (Reprinted with permission from The Apprentice School, Newport News Shipbuilding.)

PERSONAL DATA

NAME — LAST	FIRST	MIDDLE	SOCIAL SECURITY NUMBER	DATE OF BIRTH

	STREET	CITY	STATE	ZIP CODE	AREA CODE PHONE NUMBER
LOCAL ADDRESS					
HOME ADDRESS					

HEIGHT	WEIGHT	ARE YOU A U S CITIZEN? YES NO

NAME OF PERSON TO NOTIFY IN CASE OF EMERGENCY	ADDRESS	PHONE	RELATIONSHIP

NAMES OF RELATIVES PRESENTLY EMPLOYED AT THIS COMPANY	NAME	RELATIONSHIP	DEPARTMENT	SOC SEC NO

DO YOU PRESENTLY WORK FOR THIS COMPANY?	HAVE YOU PREVIOUSLY WORKED FOR THIS COMPANY?	
NO YES DEPT ________	YES NO DEPT ________	MO YR DATE LEFT ________

APPRENTICE CRAFT DESIRED

FIRST CHOICE	SECOND CHOICE	WOULD YOU ACCEPT ANOTHER CRAFT? YES NO

SCHOLASTIC RECORD

HIGH SCHOOL ATTENDED	ADDRESS OF HIGH SCHOOL (CITY AND STATE)	HAVE YOU GRADUATED? YES NO	DATE OF GRADUATION MO YR

HAVE YOU ATTENDED AN EDUCATIONAL INSTITUTION ABOVE HIGH SCHOOL LEVEL? YES LIST BELOW NO

NAME AND ADDRESS OF SCHOOL	COURSE OF STUDY	DATES ATTENDED FROM	TO	DURATION	DEGREE CERT
		FROM	TO	QTRS ____ SEMS ____ YRS ____	
		FROM	TO	QTRS ____ SEMS ____ YRS ____	

CHECK THE COURSES YOU HAVE COMPLETED IN HIGH SCHOOL OR COLLEGE. INCLUDE COURSES IN WHICH YOU ARE CURRENTLY ENROLLED. SHOW THE NUMBER OF CREDITS OR HOURS RECEIVED FOR EACH.		CREDITS		CREDITS
	ALGEBRA (1ST YEAR)	____	MECHANICAL DRAWING	____
	ALGEBRA (2ND YEAR)	____	VOCATIONAL TECHNICAL	____
	GEOMETRY	____	ADVANCED MATH	____
	PHYSICS	____	OTHERS (LIST)	____
	CHEMISTRY	____		____

LIST ANY SCHOOL EXTRACURRICULAR ACTIVITIES AND THE NUMBER OF YEARS YOU PARTICIPATED IN EACH (*Athletics, Clubs, Organizations, Etc.*)

PERSONAL REFERENCES

LIST THREE

	NAME	STREET	CITY	STATE	ZIP
1					
2					
3					

MILITARY STATUS

BRANCH OF SERVICE	TOTAL PERIOD OF ACTIVE DUTY	FINAL RANK	TYPE OF DISCHARGE
	FROM TO		HONORABLE GENERAL OTHER
SPECIAL TRAINING OR SCHOOLS IN SERVICE			MEMBER OF RESERVE YES NO ACTIVE INACTIVE

EMPLOYMENT HISTORY

FROM	TO	NAME AND ADDRESS OF COMPANY	POSITION	RATE	REASON FOR TERMINATION

Figure 7.3 (continued) Newport News Shipbuilding "application to apprentice school." (Reprinted with permission from The Apprentice School, Newport News Shipbuilding.)

SUPPLEMENTAL INFORMATION

1 ARE YOU NOW OR HAVE YOU EVER BEEN A MEMBER OF ANY ORGANIZATION WHICH HAS BEEN DESIGNATED BY THE U.S. ATTORNEY GENERAL AS HAVING INTERESTS IN CONFLICT WITH THOSE OF THE UNITED STATES? IF YES, STATE NAME OF ORGANIZATION, DATES AND EXTENT OF MEMBERSHIP

YES NO

2 HAVE YOU EVER HAD A SECURITY CLEARANCE DENIED, REVOKED OR SUSPENDED BY THE U.S. GOVERNEMENT? DETAILS

YES NO

3 ARE YOU A REPRESENTATIVE OF A FOREIGN INTEREST?
LIST FOREIGN INTEREST

YES NO

4 DO YOU OR YOUR (WIFE HUSBAND) HAVE RELATIVES LIVING IN A COMMUNIST BLOC COUNTRY?
LIST NAME, RELATIONSHIP, AND COUNTRY

YES NO

5 HAVE YOU EVER BEEN ARRESTED, CHARGED OR COURT MARTIALED FOR WHICH YOU WERE CONVICTED, FINED OR FORFEITED BOND?
IF YES, LIST BELOW

YES NO

DATE	CHARGE	PLACE OF ARREST (CITY)	ACTION TAKEN

ATTACH ADDITIONAL SHEETS IF MORE SPACE IS NEEDED

I CONSENT TO TAKING THE PRE-EMPLOYMENT PHYSICAL EXAMINATION AND SUCH FUTURE PHYSICAL EXAMINATIONS AS MAY BE REQUIRED BY THE COMPANY. I AGREE TO WEAR OR USE PROTECTIVE CLOTHING OR DEVICES AS REQUIRED BY THE COMPANY AND TO COMPLY WITH THE SAFETY RULES

I UNDERSTAND THAT MEMBERSHIP IN THE APPRENTICE SCHOOL STUDENTS' ASSOCIATION (ASSA) IS REQUIRED OF ALL APPRENTICES. AS OF JANUARY 1, 1990, DUES WERE $2.25 PER WEEK. THE LEVEL OF DUES MAY BE INCREASED IN ACCORDANCE WITH THE STUDENTS' ASSOCIATION CONSTITUTION. IF ACCEPTED, I AGREE TO PAY THE LEVEL OF DUES AUTHORIZED

I VOLUNTARILY GIVE NEWPORT NEWS SHIPBUILDING THE RIGHT TO MAKE A THOROUGH INVESTIGATION OF MY PAST EMPLOYMENT AND ACTIVITIES, AGREE TO COOPERATE IN SUCH INVESTIGATION, AND RELEASE FROM ALL LIABILITY OR RESPONSIBILITY ALL PERSONS, OR CORPORATIONS SUPPLYING SUCH INFORMATION

I AGREE THAT THE ENTIRE CONTENTS OF THIS APPLICATION FORM, AS WELL AS THE REPORT OF ANY SUCH EXAMINATION, MAY BE USED BY THE COMPANY IN WHATEVER MANNER IT MAY WISH.

IF EMPLOYED BY THE COMPANY, I UNDERSTAND THAT SUCH EMPLOYMENT IS SUBJECT TO THE SECURITY POLICIES OF THE COMPANY. I FURTHER UNDERSTAND THAT IF THE POSITION FOR WHICH I AM HIRED REQUIRES ACCESS TO CLASSIFIED INFORMATION AND I AM NOT ABLE TO OBTAIN A SECURITY CLEARANCE, I WILL NOT BE ALLOWED TO WORK IN THIS POSITION. MY EMPLOYMENT WITH THE COMPANY IN A POSITION NOT REQUIRING A SECURITY CLEARANCE DEPENDS UPON THE EXISTENCE OF SUCH A POSITION FOR WHICH I AM QUALIFIED

I FURTHER UNDERSTAND THAT ANY FALSE ANSWERS OR STATEMENT MADE BY ME ON THIS APPLICATION OR ANY SUPPLEMENT THERETO, OR IN CONNECTION WITH THE ABOVE MENTIONED INVESTIGATION, WILL BE SUFFICIENT FOR IMMEDIATE DISCHARGE

NICKNAMES OR ALIAS - MAIDEN NAME IF MARRIED WOMAN ____ APPLICANT'S SIGNATURE ____ DATE ____

ADDITIONAL COMMENTS OR INFORMATION YOU WISH TO SUPPLY IN SUPPORT OF YOUR APPLICATION

DO NOT WRITE BELOW THIS LINE

		ACHIEVEMENT		
APPLICATION FEE	ALGEBRA I		LACKS CREDIT	
APPLICATION CHECKED	ALGEBRA II		INT. DATE	
OFFICIAL H.S. TRANSCRIPT	GEOMETRY		PHYSICAL DATE	
OFFICIAL POST-SECONDARY TRANSCRIPT	CHEMISTRY		OFFER MADE	
	PHYSICS		OFFER ACCEPTED	
	VOCATIONAL / TECHNICAL		STARTING DATE	
	MECHANICAL DRAWING		TRADE	
	ADVANCED MATHEMATICS		STARTING RATE	

Figure 7.3 (continued) Newport News Shipbuilding "application to apprentice school." (Reprinted with permission from The Apprentice School, Newport News Shipbuilding.)

THE APPRENTICE SCHOOL
NEWPORT NEWS SHIPBUILDING
STATEMENT OF REFERENCE FOR APPLICANT

Part I, to be completed by applicant

To The Apprentice School Admissions Office:

I hereby authorize inclusion of this statement of reference in my admissions file and consideration of its contents in making an admissions decision. I acknowledge this information to be within the scope of the Company's investigation of my past employment and activities, which I authorized on my application form.

Please check the appropriate box:

I ( ) waive ( ) do not waive my right to see the completed statement of reference.

Name (Printed)	Date	Social Security No.	Signature

PART II, to be completed by reference

1. How long have you known the applicant?

   ____ less than 3 years
   ____ 3 to 10 years
   ____ 10 years or longer

2. What is your relationship to the applicant?

   ____ Relative
   ____ Teacher/former teacher
   ____ Supervisor/former superv.
   ____ Neighbor/friend of family
   ____ Personal friend
   ____ Other

3. Can the applicant successfully complete technical course work beyond high school level?

   ____ Yes, more than capable
   ____ Yes, able to do so
   ____ Unsure, may have difficulty
   ____ Don't know
   Other comments:

4. Can the applicant succeed in craft training and adapt to a production environment?

   ____ Yes, has good mechanical aptitude
   ____ Yes, able to do so
   ____ Unsure, may have difficulty
   ____ No, lacks mechanical aptitude
   ____ Don't know
   Other comments:

Figure 7.4 Newport News "reference form." (Reprinted with permission from The Apprentice School, Newport News Shipbuilding.)

5. What do you know about the applicant's motivation and other job-related personal characteristics?

_____ Highly motivated, with exceptional personal characteristics
_____ Well motivated, with good personal characteristics
_____ Unsure, may be lacking in motivation or good personal characteristics
_____ Lacks motivation or has shown poor personal characteristics
_____ Don't know
Other comments:

6. What do you know about the applicant's attendance trends in school and/or work?

_____ Exceptional attendance in school and/or work
_____ Satisfactory attendance in school and/or work
_____ Unsure, may be lacking in good attendance habits
_____ Has shown poor attendance in school and/or work
_____ Don't know
_____ Other comments:

7. What is the probability that the applicant will remain with the Company upon completion of apprenticeship?

_____ Excellent
_____ Good
_____ Fair
_____ Poor
_____ Don't know
Other comments:

8. If selected, the applicant will be working on defense contracts and will be required to obtain a security clearance. Can the applicant be trusted with information important to national security?

_____ Yes
_____ No
_____ Don't know
Other comments:

9. What other information relevant to consideration of this applicant do you wish to provide?

Name (Printed)   Date   Signature of Person Recommending

Telephone Number (days)   Telephone Number (evenings/weekends)

JHH/FOR/sral

Figure 7.4 (continued) Newport News "reference form." (Reprinted with permission from The Apprentice School, Newport News Shipbuilding.)

TESTING

Candidate testing is done to determine the state of readiness of each candidate. Testing prior to acceptance into the program includes:

- academic achievement (math, science, communications)
- mechanical skills dexterity assessment
- physical agility testing
- vocational aptitude assessment
- socialization skills

While not every program will assess in each of these areas, these are generally the areas of concern of many, if not most, apprenticeship program sponsors. Table 7.1 lists the commonly available tests in each of the areas and provides some details about tests that are available in each of these categories.

Academic Achievement Testing

Where specific academic skills are necessary as a prerequisite to commencement of training, the admissions committee will wish to determine the extent to which these skills are present in the candidate apprentice. Most programs are concerned with the mathematics skills (perhaps algebra, trigonometry, precalculus) and science areas.

Very often, apprenticeship admissions committees use tests that have been developed by their own trade associations. For instance, the National Tooling and Machining Association uses its own test for these purposes. The test concentrates on the mathematics and physical sciences essential to apprenticeship in the machine trades.

Mechanical Skills Dexterity Assessment

In some areas such as automotive, welding, dental technology, or watchmaking, motor skills are essential. The committee might wish to make this a form of assessment. Either a customized assessment or a standardized test can be used for this purpose. Remember, in some cases, the scoring of the test must be done by a trained person, and such arrangements might need to be made prior to test administration.

Physical Agility Testing

This form of assessment is very common in some areas of training such as in the public services (police, fire, military). For example, in the case of

TABLE 7.1. **Apprentice Entrance Testing.**

Type of Test	Publisher/Source	Content
Academic Achievement		
California Achievement Test		Tests reading, math, and language
Stanford Achievement Test		Tests measure vocabulary, spelling, and work study skills
Woodcock-Johnson Psycho-Educational Battery		Tests reading, math, written language, and knowledge of social studies, science, etc.
Mechanical Skills		
National Occupational Competency Testing	NOCTI, Atlanta, GA	Individual trade tests requiring hands-on performance
Physical Agility		
	Rancho Santiago College, Santa Ana, CA, Attn. Fire Service Program	Physical agility test battery, specifically designed for fire service personnel
Vocational Aptitude		
Armed Services Vocational Aptitude Battery	U.S. Department of Defense, Washington, D.C. 20402	Nine tests that yield three academic scales and four occupational scales; tests arithmetic, tool knowledge, space relations, mechanical comprehension, shop, automotive, and electronics information
Strong Vocational Interest Bank	Stanford University Press, Stanford, CA	Items that assist to help the candidate understand one's career interests; jobs, activities, school subjects, etc., are focused upon to help in the prediction

TABLE 7.1. (continued).

Type of Test	Publisher/Source	Content
Vocational Aptitude *(continued)*		
General Aptitude Test Battery	U.S. Employment Service, Washington, D.C. 20210	Eight paper and pencil and four apparatus tests; measure nine abilities, including spatial aptitude, finger dexterity, manual dexterity
Flanagan Aptitude Classification Test	Science Research Associates, 155 Wacker Drive, Chicago, IL 60606	Sixteen subtests that collectively measure behaviors critical to job performance: ingenuity, assembly, coding, memory, precision
Socialization		
Guilford-Zimmerman Temperament Survey	Sheridan Psychological Services, Orange, CA 92667	Measures ten traits: restraint, ascendance, sociability, emotional stability
Sixteen Personality Factor (16PF)	Institute for Personality & Ability Testing, Champaign, IL 61820	Measures sixteen personality factors related to occupational fitness projections
Interest Inventory		
Career Assessment Inventory	National Computer Systems, Minneapolis, MN	45-minute test three types of scales 1. general occup. 2. basic interest 3. occupational Primarily used with noncollege-bound students
Brainard Occupational Preference Inventory	The Psychological Corp., San Antonio, TX	Scores yield preference for commercial, mechanical, professional, aesthetic, scientific fields and agriculture (male only) personal service (female only)

the fire science technology programs at Rancho Santiago College in California, the college provides this testing as a service to the employers seeking trainees. Costs of the assessment might be borne by the candidate or the employer.

Vocational Aptitude/Interest Assessment

A very basic form of assessment used in some trade training programs is the simple vocational aptitude test. This form of assessment seeks to determine whether a person has the basic cognitive tools and interests necessary to master particular kinds of trades. It is a useful tool if part of your mission is career guidance. In fact, some batteries are good for counseling students prior to their making decisions about careers and making applications to various training programs.

Socialization Skills

Lastly, some programs are concerned with the degree to which students are ready to work in certain kinds of social situations. Working with people in certain circumstances is a very important skill. Interview committees will want an empirical basis for making decisions about social skills in separating out successful candidates for apprenticeship.

With all of the information collected, the admissions committee will now review the information and come to a decision regarding which of the candidates will be offered apprenticeship positions. In larger programs, part of the decision making will be identifying appropriate work sites and employers for each of the candidates accepted into the program.

EMPLOYER RECRUITMENT

Another part of the educational partner's role and responsibility in apprenticeship program administration is to ensure that the work sites to which the apprentices are assigned meet the requirements of educational suitability and appropriateness. For a work site to be educationally suitable and appropriate, it must be safe, properly equipped and maintained, and satisfactorily located within the community. Additionally, the employer and staff must be willing to accept the apprentice as a learner and worker and provide appropriate training and supervision—in partnership with the educational institution and others in the partnership. Following are some of the parameters that need to be addressed in considering a work site.

RECRUITING EMPLOYERS

Employers who are willing to participate in apprenticeship need to be continually identified for an apprenticeship program to be developed adequately. Where do we find such employers? Again, the educational partner is instrumental in identifying employers, and the educational institution is well positioned within a community to make such connections. Discussed next is where some of these connections can be made.

Get the Word Out at Advisory Committee Meetings

All occupational education programs are guided by advisory committees. This is a primary place to recruit both employers and apprentices. The educational partners should make their advisory committee members salespeople for apprenticeship. They should ask them to open the doors into firms in the community that are appropriate for placements. If necessary, they should charge each member with producing a list of six firms and principals who should be contacted about providing a work site. The educational partner, possibly together with employers who do sponsor apprentices, should then contact each firm and principal. In addition to securing that firm's participation, they should strive to come away with a list of at least three more firms that are potential participants.

Additionally, the advisory committee should be able to provide the names of industry associations and contacts connected to the industry or occupation under consideration. Again, initiate these contacts and recruit the assistance of these people to identify firms and principals to speak to as well.

Meet Firm Principals and Market the Program

The key person responsible for apprenticeship program development should consider preparing some literature describing the apprenticeship program, the organization sponsoring the program (e.g., if a foundation or other kind of organization is involved), and/or the participating vocational-technical school and/or community college. Make sure the literature describes the role(s) and requirements of each organization, including the firm, and the value of apprenticeship. If a foundation concept has been decided upon to underwrite the apprenticeship and if the apprentice will actually work for the foundation, describe how the billing process will work for the apprentice's services. If the apprentice will be employed by the firm, discuss the requirements for a progressive wage schedule. Discuss the work process schedule and the firm's ability to comply with such a schedule, given its product or service lines. Discuss each of these points with the firm's principal, stressing the value to the firm participating. Give the firm time to consider participating. Make a follow-up appointment to return. Figure 7.5 is a form used by a state department of labor for an employer to apply to serve as a work-site sponsor.

EMPLOYER AND WORK-SITE SCREENING

WORK-SITE SELECTION

While visiting each of the firms, apprenticeship program representatives will need to gather enough information on the firm to make decisions about each firm's appropriateness to serve as a work site.

Developing Work-Site Standards

What will you look for and what sort of information will you need? Consider:

(1) Work-site supervisors as instructors
(2) Equipment availability and condition
(3) Safety
(4) Level of business activity and kind of work being done by firm; comprehensiveness of work experience for the apprentice
(5) Overall experience and reputation of firm; quality of working relationship and willingness to comply with work schedule
(6) Location of firm
(7) Possibility of work layoffs

Figure 7.6 is a form used by MechTech, Inc. to sign up employers.

Work-Site Supervisors as Instructors

A very important consideration is the philosophy and attitude of the employer and the employer's staff to provide instruction and support to an apprentice. The supervisors with whom the apprentice will be working must believe in apprenticeship as a training process. These people must be willing to provide patience, support, and guidance to the apprentice. They must be willing to work with you and the organization as a partner in the process. If it is found that any one of these items is not present, consider not involving

AT 2 SUP

CONNECTICUT DEPARTMENT OF LABOR APPLICATION FOR APPRENTICESHIP SPONSOR

GENERAL INFORMATION:

1. NAME OF FIRM ______

D/B/A NAME, IF ANY ______

MAILING ADDRESS ______ ZIP ______ FAX NO. ( ) ______

ACTUAL LOCATION ______ ZIP ______ PHONE NO. ( ) ______

CITY ______ COUNTY ______ STATE ______

2. TYPE OF FIRM (Check [✓] only one) ______ CORPORATION ______ PARTNERSHIP ______ Sole Proprietorship

3. HOW MANY YEARS HAS THE FIRM BEEN IN BUSINESS? ______ UNDER THE SAME NAME? ______

4. WITHIN THE PAST FIVE YEARS HAS THE FIRM, ANY AFFILIATE, (INCLUDING ANY CONTRACTOR OF RECORD) ANY PREDECESSOR COMPANY OR ENTITY, OWNER OF 5.0% OR MORE OF THE FIRM'S SHARES, DIRECTOR, OFFICER, PARTNER OR PROPRIETOR BEEN THE SUBJECT OF: (Check any that apply and explain under sponsor remarks. It is imperative that a full explanation of the circumstances relating to a "yes" statement be submitted to ensure an objective evaluation by the Department. Attach additional pages if necessary.)

A. a judgement of conviction for any business-related conduct constituting a crime under state or federal law? ______ yes, ______ no
B. a currently pending indictment for any business-related conduct constituting a crime under state or federal law? ______ yes, ______ no
C. a grant of immunity for any business-related conduct constituting a crime under state or federal law? ______ yes, ______ no
D. any federal determination of a violation of any labor law or regulation? ______ yes, ______ no
E. any OSHA violation that was categorized as willful, repeat, failure to abate, or was based on retaliating against an employee for filing a safety or health complaint? ______ yes, ______ no
F. any state determination of a violation of any labor law or regulation? ______ yes, ______ no Public work violation? ______ yes, ______ no Was this violation willful? ______ yes, ______ no
G. a consent order with the Connecticut Department of Environmental Protection, or a federal or state enforcement determination involving a construction-related violation of federal or state environmental laws? ______ yes, ______ no
H. a debarment from federal contracts for violation of the Davis-Bacon Act, 49 Stat 101(1931), 40 USC 276a-2 or pending enforcement proceeding for same. ______ yes, ______ no
I. a debarment from state contracts for violation of Connecticut's prevailing wage law pursuant to Conn. Gen. Stat. Section 31-53a of the General Statutes, or pending enforcement proceeding for the same? ______ yes, ______ no
J. a debarment or suspension for violation of any other state prevailing wage law or pending enforcement proceeding for same. ______ yes, ______ no
K. rejection of any bid or proposed subcontract or general contract for lack of responsibility pursuant to state law? ______ yes, ______ no
L. any determination by the Connecticut Department of Consumer Protection related to violations of a state occupational licensing statute or regulation or pending enforcement proceeding regarding the same. ______ yes, ______ no
M. a consent order entered into with the Connecticut Department of Consumer Protection or any other State or Federal government agency. ______ yes, ______ no

- None of the Above ______
- Does firm have an active Job Order with the Department for Journeypersons? ______ yes, ______ no
- Does firm have an active Job Order with the Department for Apprentices? ______ yes, ______ no

Figure 7.5 Connecticut DOL sponsor form. (Reprinted with permission from the Connecticut Dept. of Labor.)

SPONSOR'S REMARKS:

For Office Use Only

Received ESU ________
Contents Reviewed and Verified by ____
Received C.O. ________
Approval Yes ____ No ____
Initials ______________

CERTIFICATION: The undersigned acknowledges that this questionnaire is submitted for the express purpose of inducing the Connecticut Department of Labor to authorize the hiring of apprentices under its state apprenticeship program pursuant to Section 31-51d-1-12 of the Regulations of Connecticut State Agencies. Applicant acknowledges that the Department may, in its discretion, determine the truth and accuracy of all statements made herein. Applicant further acknowledges that intentional submission of false or misleading information in this application may constitute reasonable cause for invalidating the applicants apprenticeship program pursuant to Section 31-51d-3a of the Regulations of Connecticut State Agencies. Applicant states and certifies under penalty of law (Conn. Gen. Stat. Section 53a-175-Class A Misdemeanor) that the information submitted in this questionnaire and any attached pages is true, to the best of his or her knowledge.

______________________	______________	______________________
Signature of Officer	Date	Contractor of Record
Printed or Typed Name of Officer	Title	Printed or Typed Name of Contractor of Record

Figure 7.5 (continued) Connecticut DOL sponsor form. (Reprinted with permission from the Connecticut Dept. of Labor.)

MechTech, Inc.

Uniting Industry and Education for Metalworking Apprenticeships

70 Glen Road, Cranston, RI 02920
TEL: 401-461-8605 • FAX: 401-461-1119

LETTER OF INTENT FOR MECHTECH MEMBER COMPANIES

The undersigned company endorses the MECHTECH program as outlined by its representative and would like to join in the training efforts for its apprentice/student employees. We also understand and support its stated Non-Hiring Agreement of any apprentice in the program.

Scheduling will be subject to apprentice availability and is at the discretion of the Administrator based on the needs of the apprentice. It is also our understanding that we may temporarily suspend our company's participation in a specific training term by notifying the Administrator prior to apprentice scheduling. This does not jeopardize our future involvement with the program.

Our company fully understands the MechTech rotation concept of these apprentices and is committed to the training goals the MechTech Board of Directors is achieving with the program. As active participants, our company will TRAIN MechTech apprentices in our machining or toolroom facilities, offer them practical work experiences, and inspire safe and diligent work habits for the apprentices.

MechTech is a non-profit corporation supported by the industry it serves and a recognized member in good standing of the National Tooling & Machining Association.

Our company will be billed weekly for all hours worked by the MechTech apprentice and we agree to pay invoices according to the NET 10 DAYS TERMS.

Company: Date:

Authorized Signature: Telephone: ( )

Mailing Address: Fax: ( )

City: State: Zip Code:

"We're building America's workforce to be the best in the world."

A Non-Profit Corporation

Maryland • New Hampshire • Massachusetts • North Carolina

Figure 7.6 MechTech "letter of employer intent." (Reprinted with permission from MechTech, Inc.)

this firm in the program because you will have a very difficult road ahead of you in making the apprenticeship meaningful for your trainee.

Review employer and staff qualifications also. In considering the employer and his staff as work-site supervisors, it will be necessary to establish each of their qualifications in the trade or occupation. Ask these questions directly. If they have superior qualifications, they will be proud to offer them to you.

Equipment Availability and Condition

For trades and occupations that are equipment intensive, the availability and condition of equipment are important. Equipment that is representative of the state of the art of the industry must be on hand. The apprentice must be exposed to the kinds of equipment that he/she will encounter when a journeyman. Adequate equipment must be on hand for the apprentice to use and gain experience on, and the equipment must be in good condition and safe. Where adequate equipment is not on hand, but the work site is otherwise very desirable, perhaps a situation where the apprentice is rotated to another shop or employer for a portion of the indenture might be appropriate and a good solution.

Safety

A safe working environment is essential to a good apprenticeship. As the educational partner, you have a legal liability to ensure that your apprentice is working in a safe environment. Check out the safety record of the employer through the appropriate agency in your state. Inquire about the employer's safety program within the firm. Look for indicators of safety in the plant—fire extinguishers, sprinkler system, first aid equipment, designated safety officer, and so on.

Level of Business Activity and Kind of Work Being Done by Firm

How comprehensive will the work experience be for the apprentice? Does the firm have enough work backlog to be able to predict that the apprentice will be able to be exposed to the entire work schedule? Is the firm engaged in the overall aspects of the trade or occupation? Most probably, a firm will not practice all aspects of a trade or occupation. This is one reason why a system of rotations is preferable, whereby the apprentice actually is employed by the educational foundation and rotated through those shops that practice particular aspects of a trade. Alternatively, you will need to make arrangements for the apprentice to gain experiences otherwise not able to be achieved in a particular shop.

Overall Reputation of Firm

What do you know about the principals of a firm? What might the quality of the working relationship be? Will there be a mutual willingness to comply with a work schedule? Will the apprentice's experience be recognized by the industry as a result of working for a particular firm and supervisor? A firm's reputation should be considered in making the decision to place the apprentice.

Location of Firm

Of course, the location of the work site is of importance as well. The apprentice must be able to travel to the work site. This is somewhat fundamental, but in the case of preapprentices, this is absolutely a consideration.

Possibility of Work Layoffs

In these days of economic uncertainty, the possibility of work slowdowns needs to be factored into work-site selection. Again, one solution to this dilemma is a system of rotations through a number of shops or to ensure that the work backlog in a particular placement is sufficient to ensure that the apprentice will get the needed training. In any event, planning is a necessity in this regard. Figure 7.7 (p. 178) is a firm's company rating form.

TRAINEE PLACEMENT AND COUNSELING

Following this, the indenture agreement is signed. Once the apprentice is placed on the job, the role of the educator in this partnership is usually to provide support to the training process. This support comes in the form of problem resolution if and when problems arise on the job. It also involves interfacing with the employer and the employer's apprenticeship supervisor on a regular basis to ensure that problems do not arise or are solved quickly before they get out of control. Communication is the key, as was already discussed in detail in Chapter 2.

The employer and supervisor should know how to get in touch with the

COMPANY RATING REPORT

TRAINING SITE: ______________________ DATE: ____________________

YOUR NAME: ____

Rate your experience or opinion of this training site for the most recent rotation.

1 = Below Average **2 = Average** **3 = Above Average** **4 = Excellent**

I. EQUIPMENT:

A. Condition (ability to hold tolerances) ________
B. Capacity (correct type for job perfomance) ________
C. Availability (did you have to wait to do jobs) ________
D. Shop tools (adequate, condition) ________

II. SAFETY:

A. Equipment available and in use ________
B. Instructions on safety procedures ________
C. General housekeeping, organization, clean shop ________

III. INSTRUCTIONS:

A. Instructions provided to do assignments ________
B. Regarding time allotted to perform jobs ________
C. Organization of work assignments ________
D. Informal progress and performance reviews ________

IV. OVERALL EXPERIENCE AT THIS TRAINING SITE:

A. Variety of learning and doing ________
B. Opportunity for individual performance ________
C. Degree of job related stress ________
D. Other personnel ________

V. COMMENTS:

____
____
____
____
____
____
____

Figure 7.7 MechTech "company rating form." (Reprinted with permission from MechTech, Inc.)

educational point of contact quickly when necessary. They should understand that it is desirable and okay to call upon the school or college instructor for briefings and updates on apprentice progress. They should also understand that the educational partner needs regular reports and formal evaluations of apprentice progress.

Likewise, the apprentice should understand how to contact the instructor when assistance is needed to resolve a problem. The apprentice should understand that it is okay to ask for help in clarifying a problem such as enforcing the agreement to provide a structured training program in accordance with the schedule of work when an employer forgets that such exists.

APPRENTICE HANDBOOK

One useful method for conveying to the apprentice and employer the "rules of the game" is an apprentice handbook. Used by most all major sponsors of apprenticeship, a handbook spells out useful and pertinent information on the following subjects:

(1) Work hours and conditions
(2) Benefits
(3) Resolving work-related problems
(4) Time sheets and work records
(5) Apprenticeship program requirements
(6) School program and requirements
- advising and counseling
- registration
- academic standards and grades
- attendance in classes
- academic behavior
- degree requirements
- applying for graduation

TRAINEE EVALUATION ON-THE-JOB AND IN THE CLASSROOM

Part of the responsibility of the educational partner is to coordinate the overall evaluation of the apprentice. This happens by ensuring that the on-the-job supervisor knows what kinds of evaluations are expected, knows the forms to be used, and knows the schedule for submitting them. All of these items become the educational partner's responsibility for coordinating. The procedures for designing the evaluation instruments (tests, etc.) will be covered in Chapter 9.

EMPLOYER PREPARATION

ESTABLISHING AN UNDERSTANDING OF THE REQUIRED WORK SCHEDULE

Once a decision is made to place an apprentice with a particular employer, it is essential to make sure that the supervisor of the apprentice understands the work processes that the apprentice will be required to master. Leave a copy of the schedule of work processes with the supervisor. Also, establish an understanding of the method by which the apprentice will be evaluated in the work processes (more about evaluation in Chapter 9).

SUPERVISOR TRAINING

If the apprenticeship program develops into a large-scale operation, at some point you will wish to commence formal training of supervisors as instructors. Some industries have such programs, such as the fire service and nuclear power industry, where formal training exists for work-site trainers. Programs exist whereby you can train these supervisors through short courses, making them more prepared to formally instruct apprentices. More about this is given in Chapter 9.

SUMMARY

This chapter presented necessary information about recruiting, screening, and testing apprentices. It also discussed identifying and screening employers for cooperative apprenticeship. When students and employers are properly matched for cooperative apprenticeship training, wonderful chemistry develops, which culminates in the best skilled training possible for the student. Simultaneously, the employer benefits from cultivating a long-term skilled worker.

REFERENCE

Cantor, J. A. (1994). *Cooperative Education and Experiential Learning*. Toronto, Canada: Wall & Emerson, Inc.

CHAPTER 8

Related Classroom Education and Educational Support

TRADITIONALLY the role of the educational institution in apprenticeship is in the area of related classroom training; however, many educational institutions go far and beyond simply providing related classroom education and training. Many vocational-technical schools and community colleges also provide preadmissions testing and counseling services. Some provide remedial education classes, and some offer college-level credit for related coursework and lifelong learning experiences through the community college. Many community colleges offer specially designed degree programs for apprentices. Some educational institutions also train employer-supervisors as work-site and classroom instructors. This chapter provides insights into the various kinds of services the Voc-Tech and community college can provide to the educational partnership for apprenticeship training (see Figure 8.1). Other important topics will include classroom logistics and training the on-the-job supervisor. Let's look at the following areas:

- preadmissions testing and ongoing counseling
- designing and developing related training classroom courses
- designing and developing Associate degree programs for apprentices
- credit for prior learning
- classroom logistics
- instructor development

PREADMISSIONS TESTING AND ONGOING COUNSELING

A very essential service that the Voc-Tech and community college can offer an employer and joint apprenticeship committee is preadmissions testing and counseling. This applies to both the apprenticeship and collegiate experience. While testing is covered in Chapter 9, a few words are in order here.

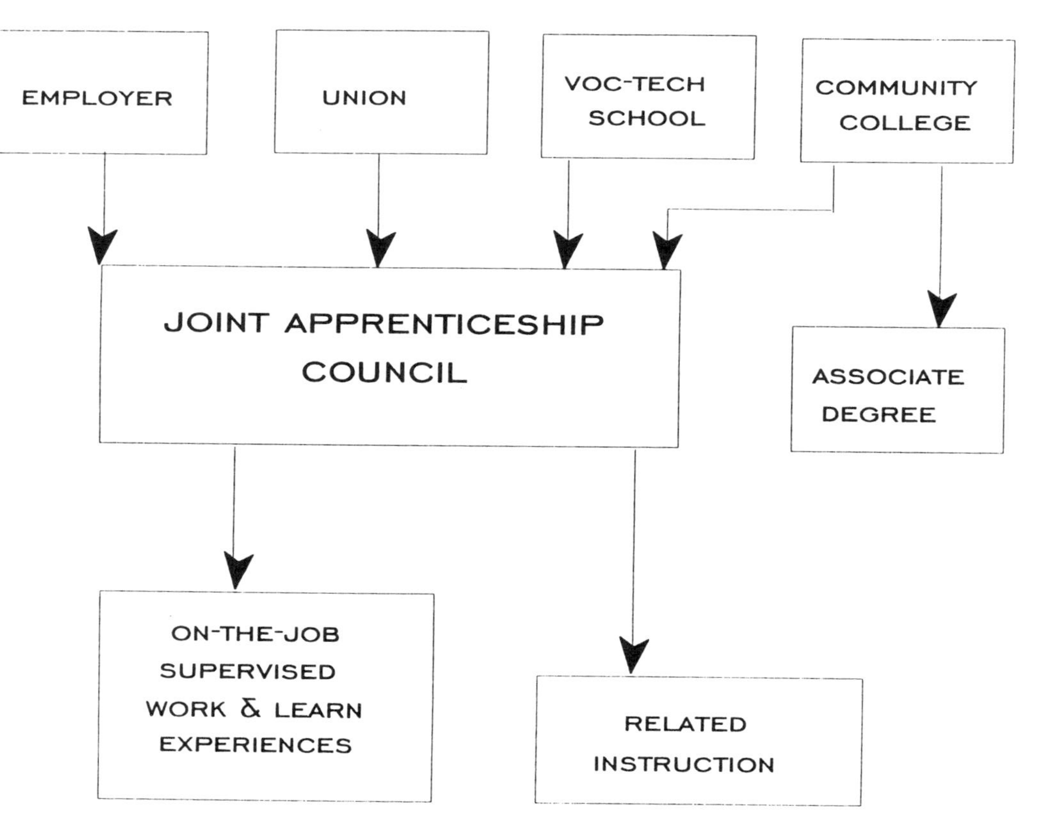

Figure 8.1 A partnership for education and training.

The apprentice will need a minimum level of basic proficiency in computational and communications skills. These basic entry-level skills must be assessed prior to acceptance to an apprenticeship. Where some remedial work is in order, the educational institution is in the position to provide it to the apprentice. The apprentice should meet the required skill levels in these various areas in order to proceed with the apprenticeship courses, as well as the Associate degree requirements.

PRIVACY OF INFORMATION RELEASES

Since both the educational institution counseling staff and JAC will need access to certain private data and information, including educational records, test results, course completion grades, and so on, it is advisable to have the apprentice complete a release of information form (see Figure 8.2, p. 184). Once the form is on file, educational test and course completion results can be released to the JAC as well.

DESIGNING AND DEVELOPING RELATED TRAINING CLASSROOM COURSES

Perhaps the most common activity that the participating vocational-technical school and community college offer to employers in support of apprenticeship is the coursework to satisfy the required related training. Most often, these courses are of the noncredit nature, offered as adult education services through the evening division in the Voc-Tech or the continuing education division in the community college. For instance, Daytona Beach Community College (DBCC) states in their 1994–1995 catalog:

> DBCC provides instruction under the direction of industry apprenticeship committees which sponsor the apprenticeship programs. These programs are registered with the U.S. Department of Labor's Bureau of Apprenticeship Training or the state of Florida's Department of Labor and Employment Security. Students who complete a program successfully will receive certificates. (p. 72)

Daytona Beach Community College offers this instructional support to electrical, plumbing, culinary, and pastry apprentices.

Montgomery College (Maryland) offers apprenticeship training support through its Homer S. Gudelsky Institute for Technical Education. It advertises:

> We provide quality related instruction. Montgomery College has served the educational needs of County business and industry for more than forty years.

MASSACHUSETTS BAY COMMUNITY COLLEGE,
GENERAL MOTORS,
AND
MINUTEMAN REGIONAL VOCATIONAL TECHNICAL SCHOOL

AUTOMOTIVE SERVICE EDUCATIONAL PROGRAM (ASEP)

PRIVACY WAIVER

Student records are kept as required by the Privacy and Confidentiality Regulations of Massachusetts Bay Community College.

As an applicant for admission, we want you to know that the College may request letters or statements of recommendation and evaluation from listed references, former employers, and educational settings. This information may be used by the Admissions Committee to assist in selecting candidates for the Automotive Service Educational Program. We would like this information to be as objective and complete as possible.

Massachusetts law provides that, after you are admitted, you have the right to inspect and copy personal data maintained on you in our College Student files except where prohibited by law or judicial order or where you waive your right of access to confidential material.

Would you please complete one (1) of the following:

A. I,________________________________, do not waive my right to inspect and copy confidential letters or statements of recommendation or evaluation obtained in connection with my application for admission.

B. I,________________________________, waive my right to inspect and copy any confidential letters or statements of recommendation or evaluation obtained by the College in connection with my application for admission. I understand, however, that I may see all other materials included in student folder and that the College, on written request, will supply me a list of persons who submitted such confidential recommendations or evaluations.

Signed:________________________________ Date:______________

Figure 8.2 Release of information form. (Reprinted with permission from Massachusetts Bay Community College.)

In the ensuing years we have earned our reputation for providing excellent training and education; our apprenticeship related instruction maintains this standard. Classroom instruction consists of a core of general science and mathematics courses followed by technical trade-related courses. All of our instructors are journeyworkers or engineers. Classes are small, insuring individualized attention. Tutoring is offered to any student requiring or requesting it.

DESIGNING RELATED INSTRUCTION

To develop instruction appropriate to a particular apprenticeship program, the educational institution should meet with firm or industry representatives and review the occupation to be taught. This review might include an existing work processes schedule or other documentation, including National Skills Standards as described in Chapter 6, V-TECHS Catalogs, and/or other existing descriptions of the occupation and its knowledge requirements—jobs and tasks. At Montgomery College, this process is done through a DACUM (Develop a Curriculum) process, which involves a committee review (business representatives, master craftspeople, educators) of these materials, systematically identifying knowledge, skills, and attitudes necessary to perform the job. From this review, the committee identifies those jobs/skills that are to be learned on the job and then separates out the knowledge that needs to be imparted in the classroom.[1]

In addition, consideration is made about those jobs/skills that should be taught in a school/college laboratory or shop. This is often the case for safety, time, or resource allocation reasons because it is more expeditious to provide initial skills exposure in a controlled environment prior to attempting the skill on the actual job.

DOL Support to the Curriculum Development Process

State departments of labor oftentimes oversee the work process schedules for standard occupational titles. The state DOL apprenticeship representative is principally responsible for keeping the work process schedule up-to-date. Usually working with the designated Voc-Tech or community college representative, the DOL representative should periodically (depending on the nature of the occupation) meet and assemble an advisory committee comprised of employers and journeyman representatives and revisit the

[1]Additional information about course and program development can be found in: Cantor, J. A. (1992). *Delivering Instruction to Adult Learners.* Toronto, Canada: Wall & Emerson, Inc.

work process schedule through a job/task analysis. From this basis, the related training can then be updated and/or modified as needed. Montgomery College conducts this review for every apprenticeship occupational area every three years.

COURSE DELIVERY

Once a decision is made about the course content, performance objectives (per discussion in Chapter 5), course descriptions, and content outlines must be developed.

Voc-Techs and community colleges offer related training in the form of courses or course modules, depending on the apprenticeship programs that they serve. Where large numbers of apprentices are to be trained, the course format will work well. At Community College of Rhode Island, a course description that appears in the MechTech, Inc. literature for apprentices reads:

> LATHE I (1 Credit) To develop in the student the ability to identify major components and accessories of the engine lathe, to understand their operating principles and transpose this understanding into basic turning operations. Lecture: 1 hour

The colleges serving the U.S. Navy's shipyard apprentices use a program of individual courses offered on a preestablished daytime schedule on or near the Navy shipyard. All apprentices in a particular year of study/apprenticeship attend classes together. Figure 8.3 displays such a program.

TEXTBOX 8.1

U.S. Navy, Naval Aviation Depot Apprenticeship Program

The U.S. Navy sponsors apprenticeship opportunities for civilian technicians in its facilities in Norfolk, Virginia. These opportunities are in several trade areas, including aircraft mechanic, aircraft electrician, painter, instrument mechanic, and sheet metal mechanic. The apprentice program follows a cooperative work-study plan, combining classroom instruction at one of several local community colleges with on-the-job training and trade theory classes conducted at the Depot facility.

Eligibility for admission includes U.S. citizenship, minimum age of eighteen or a high school graduate, good physical fitness, academic preparedness for college work, and having a vocational-technical school background. An application, high school transcripts, and high school recommendation are required.

Apprentices attend courses at either Thomas Nelson or Tidewater Community Colleges. They also attend related trades theory courses conducted by journeypersons at the Navy Depot. On-the-job portions of the apprenticeship are handled by rotation through the shops at the Depot. In addition to

SHIP REPAIR

The Program is designed to provide a multifaceted introduction to the basic skills which will allow students to embark on a successful career at the Pearl Harbor Naval Shipyard. Courses must be taken in the sequence and semester scheduled. Students must obtain a security clearance to be accepted by PHNSY for on the job training in WORK 94V.

The cost of textbooks is approximately $300.

Program Prerequisites: ENG 10, MATH 1		Associate in Science Degree Credits
First Semester		
ENG 21C	Technical Reading	3
MATH 50	Technical Math	3
BLPRT 22	Blueprint Reading & Drafting	3
WELD 19	Welding for Trades & Industry	3
IEDMS 101	Machine Shop for Industrial Education	3
SHIP 20	Intro to Cooperative Education	1
		16
Second Semester		
HLTH 31	First Aid & Safety	1
ENG 60	Technical Writing	3
PHYS 51V	Technical Physics	4
BLPRT 45	Naval Blueprint Reading	2
SHIP 30	Ship Painting	3
SHIP 31	Fabric Work	3
		16
Work Cycle: First Summer		
WORK 94V	Work Cycle	5
Third Semester		
PSYCH 197	Human Relations & the Workplace	3
SPEECH 151	Personal & Public Speech	3
SHIP 40*	Sheet Metal I	3
IEDSM 103	Sheet Metal for Industrial Education	3
ICS 120	Computer Literacy & Applications	3
		12
Fourth Semester		
SHIP 50*	Sheet Metal II	3
IEDWC 101	Hand & Portable Tools/Material & Hardware	3
IEDIE 102	Electrical Building Construction	3
General Education Elective	Group D** - Understanding World Cultures	3
Elective	Course(s) in Field of Specialization	3–4
		15–16
Work Cycle: Second Summer		
WORK 94V	Work Cycle	5
Minimum Credits Required		71–72

*Courses to be developed.

**General Education Requirements for the A.S. degree are listed under Degrees and Certificates.

Figure 8.3 Pearl Harbor Naval Shipyard apprentice program.

maintaining a satisfactory GPA from the college, students are evaluated in the trade theory and on-the-job portions of the program for continuance in the apprenticeship program.

[Adapted from *Cooperative Apprentice Training Program: Handbook.* (1990, July). Naval Aviation Depot, Norfolk, VA.]

Modularized Courses

Where smaller numbers of apprentices are to be served or where there are diverse course material requirements and multiple schedules to be served, alternative programming configurations must be established. One such alternative is the modularized approach as used by Montgomery College (MD). MC developed a sequence of modules in thirteen fifty-two-hour blocks. For instance, in the air conditioning/refrigeration program, Year 1 courses appear as:

Intro. to Math	Module CC 115	52 Hours
Communications Skills	Module CC 103	26 Hours
Basic Electricity I	Module CC 101	26 Hours
Blueprint Reading I	Module CC 102	52 Hours
Shop Math I	Module CC 106	52 Hours
	Total	208 Hours

Modularized courses permit the educational institution to offer a more flexible schedule of classes for apprentices. Because of this, an educational institution can offer any of the above classes once every few weeks, on campus or off, and commingle apprentices from any number of employers into the modules they are required to take.

Correspondence Courses

In some cases, either because of distance from the college for the apprentice or because of the limited numbers of people to be trained in a given subject at any one school location, correspondence courses (a form of distance learning) are used. Montgomery College uses this form of instruction for sprinkler fitting related training. North Dakota State College of Science offers a correspondence-related education program in a number of technical areas.

Remedial Education

Most educational institutions offer remedial coursework in basic skills areas—reading, English, arithmetic and general mathematics, and other pretechnical subjects. Educational institutions can offer these classes to the apprentice, either on- or off-campus at the job site.

DESIGNING AND DEVELOPING DUAL-ENROLLMENT ASSOCIATE DEGREE PROGRAMS

As the value of a college degree becomes more important for a technician, community colleges are exploring new and innovative ways and means to offer the education to the apprentice. Several novel approaches have emerged. In Oregon, the process for designing the Associate in Applied Science Degree program for apprentices corresponds to regional accreditation standards. Three basic steps are recommended for approaching Associate degree program development:

- identifying degree requirements
- developing the curriculum
- submitting the curriculum to a college curriculum committee for approval

IDENTIFYING DEGREE REQUIREMENTS

All community colleges have general education course requirements (math, English, history), as well as specific technical course requirements. These colleges will also have guidelines for the total number of credits required for the degree. From this total number, one will subtract the number of general education course credits to determine the remaining technical credit requirements that can be met through related education and on-the-job work credit.

Next, identify the work processes and hours required by the state's apprenticeship council or other cognizant body for the trade or occupation in question. The remaining credits can now be allocated among these work processes and divided into courses. Depending on accreditation policy, some or all of the credits may be awarded through credit by exam or other method of experiential learning. In Oregon, 25 percent of these remaining credits can be given by credit assessment, with the remaining placed on the student's record as nontranscripted course substitutions (see Figure 8.4).

DEVELOP THE CURRICULUM

Developing the curriculum involves gathering information from college apprenticeship instructors, journeymen, and industry representatives, as well as from documentation. The materials are prepared as course outlines

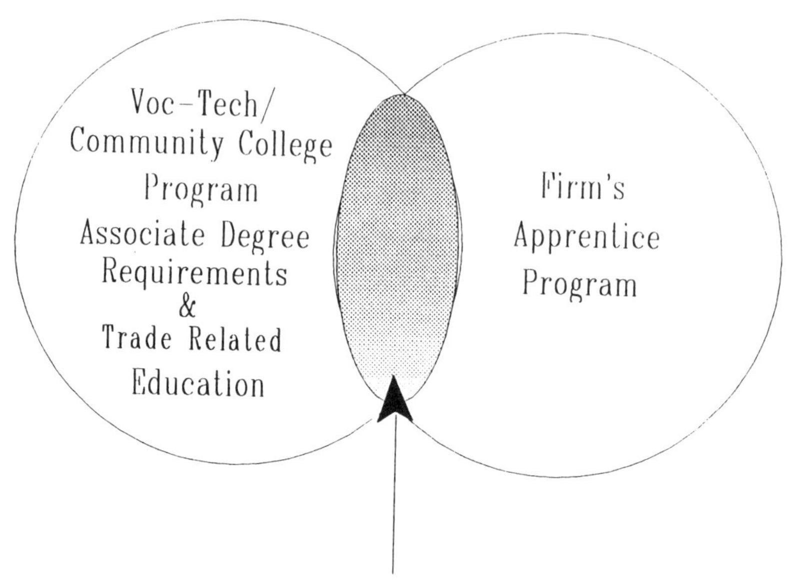

Figure 8.4 Design of instruction.

according to work processes in accordance with the state's requirements and formats. Some of these materials can be obtained from industry groups, associations, or union national organizations in already prepared Associate degree program models.

SUBMIT TO CURRICULUM COMMITTEE

Once the college's process requirements have been addressed, the package is submitted to the college curriculum committee by the appropriate advocate. Figure 8.5 is a sample format offered by the Oregon Department of Education.

PREDEVELOPED INDUSTRY–EMPLOYER/ASSOCIATION/EDUCATIONAL FOUNDATION PROGRAM MODELS

Numerous employer associations and unions have developed model dual-enrollment Associate degree apprenticeship programs. In the early 1980s, the automotive manufacturers—both those headquartered in the United States as well as importers—introduced cooperative apprenticeship programs into the industry. These programs were developed in cooperation with

Voc-Techs and community colleges and designed to bring youth from the high school into the workplace and to link the young worker into postsecondary schooling. Let's look at some of the programs.

Automotive Manufacturer Programs

Toyota Motor Corp. and Chrysler Corp. are but two manufacturers that offer an automotive technician apprenticeship training program. Toyota's program is designed as a community college Associate degree program. Toyota has specific courses and lab requirements that the college must agree

Sample Program: Apprenticeship Sheet Metal Worker

Work Processes and Approximate Hours (slightly abridged):

a.	General sheet metal work	3,100 hours
b.	HVAC, furnaces, etc.	3,200 hours
c.	Industrial sheet metal	400 hours
d.	Soldering & welding	800 hours
e.	Special installation	1,250 hours
f.	Electronic & electrical	1,250 hours
	TOTAL	10,000 hours

General Education Requirements

Course No.	Course Title	Clock Hours	Credit
WR 121	Writing 121	36	3
Mth 112	Trigonometry	48	4
HE 125	Industrial Safety	36	3
EL 115	Effective Learning	36	3
SP 111	Speech (or)	36	
SP 105	Listening	36	3
GEO 206	Geography	36	3
CS 161	Computer Science (or)	72	
6.229	Electronics	72	8

Apprenticeship Training

Course No.	Course Title	Clock Hours	Credit
9.505	General Sheet Metal I	107	10
9.507	General Sheet Metal II	100	9
9.502	HVAC, Furnaces, AC, Heat Pumps . . . I	107	10
9.508	HVAC, Furnaces, AC, Heat Pumps . . . II	100	9
9.504	Industrial Sheet Metal	27	2
9.506	Soldering & Welding	53	5
9.501	Special Installation	83	8
9.503	Electronic and Electrical	83	8
9.500	Related Training	48	6

General Ed:	27
Apprentice:	67
Total:	94

For more information contact Fred Smith at the Oregon Department of Education, (503) 378-3584 ext. 336 or Fax (503) 378-5156.

Figure 8.5 Sample program: Oregon. (Reprinted with permission from Lane Community College.)

to incorporate in order to be able to participate in this program. Toyota has a staff of field supervisors who work with the college and the participating dealer employers to ensure that the program works. Chrysler offers a similar program that has a model curriculum for Voc-Techs and community colleges to follow. The course display for this program, as initiated by Massachusetts Bay Community College, is shown in Figure 8.6.

Navistar, Inc. Apprentice Program

Navistar's program is developed with Clark State Community College, Springfield, Ohio. The program was developed in cooperation with the college and the United Auto Workers. The apprenticeship is a four-year, 8,000-hour experience for training of millwrights, electricians, and carpenters. The college is responsible for 800 hours of related classroom instruction. The college millwright program includes (Pooler, 1993):

Subject	Hours
Mathematics	250
Mechanical drawing and blueprint reading	174
Shop theory	208
Metallurgy	18
Hydraulics	18
Nontechnical subjects	100
Subjects added to above as needed	32
Total	800

LABOR–EMPLOYEE ORGANIZATION SPONSORED PROGRAMS

Several national labor unions have initiated Associate degree apprenticeship dual-enrollment models worthy of consideration. Unions such as the International Brotherhood of Electrical Workers (IBEW), International Union of Operating Engineers (IUOE), and International Association of Fire Fighters (IAFF) have promoted Associate degree dual-enrollment programs dating back over twenty years.

The Local #3—IBEW/Empire State College Program

Under the sponsorship of the Joint Industry Board of the Electrical Industry, apprentices with the IBEW Local #3, New York City can earn an entire Associate in Science degree while studying technical subjects in electrical theory and completing their on-the-job apprenticeship training

AUTOMOTIVE TECHNOLOGY: CHRYSLER (AY)

The Chrysler Apprentice program (CAP) is designed to upgrade the technical competence and professional level of the incoming dealership technician. The Program leads to an Associate in Science Degree in Automotive Service Technology. Students in the program receive "state-of-the-art" instruction on Chrysler products. The program involves academic instruction on the Mass Bay campus and lecture and laboratory focusing on Chrysler products at Assabet Valley Regional Vocational Technical School. Students are also required to work at a Chrysler dealership as part of their cooperative education. The implementation of the Chrysler (CAP) Program is a collaborative effort of Mass Bay Community College, Assabet Valley Regional Vocational Technical School, and Chrysler Corporation. The College has the academic and administrative responsibility for the program which is pursuing certification by National Automotive Technicians Education Foundation Inc. in all eight performance areas.

CURRICULUM

Term I: 12 Weeks Academic Study, 12 Weeks Cooperative Education		Credits	Credits
AY 100	Fundamentals of Automotive Technology	5	
AY 110	Automotive Electricity	4	
AY 115	Cooperative Education I		3
CT 100	Critical Thinking Strategies	2	
MA 131	Technical Mathematics	3	
	Humanities Elective	3/4	
SF 131	Speech and Oral Communication	3	—
		20/21	3
Term II: 12 Weeks Academic Study			
AY 120	Auto Electronics	3	
AY 140	Automotive Brakes	3	
AY 170	Electronic Fuel and Engine Contro	4	
CS 100	Computers & Tech	4	
EL 101	Fundamentals of Electronics	4	
EN 101	Freshman English I	3	
		21	
Term III: 8 Weeks Cooperative Education, 7 Week			
AY 125	Cooperative Education II	3	
AY 221	Heating/Air Conditioning		3
AY 230	Engine Performance		5
MG 102	Small Business Management		3
PS 260	Psychology in Business and Industry		3
	Social Science Elective	—	3
		3	17
Term IV: 12 Weeks Cooperative Education, 12 Weeks Academic Study			
AY 215	Cooperative Education III	3	
AY 245	Auto Engine Diagnosis & Repair		4
AY 250	Automatic Transmissions		4
AY 255	Manual Transmission/Drive Systems		4
AY 270	Auto Steering Suspension Systems		3
EN 102	Freshman English II	—	3
		3	18
Term V, 12 Weeks Cooperative Education			
AY 225	Cooperative Education IV	3	
		3	

Total Credits: 87/88

*If WR 100 is waived, EN 191 is recommended.
If WR 100 is required, it will fulfill the humanities elective.

Figure 8.6 Chrysler corporate program. (Reprinted with permission from Massachusetts Bay Community College.)

(Electrical Industry, 1989). The electrical theory coursework, taught by IBEW union journeymen through the New York City Public Schools, has been evaluated by Empire State College and is accepted as the equivalent of forty college-level credits.

Empire State College grants thirty-two credits towards the Associate degree to every matriculated apprentice who has completed the electrical theory courses and has passed the "M" test, the final mastery test in the apprenticeship program, which is devised jointly by the union, the joint industry board, and the college. The Associate degree requires the completion of sixty-four credits, of which at least thirty-two credits must be in liberal arts and taken at the college. The remaining eight electrical course credits are applicable towards a bachelor degree at Empire State College. The required credits for the Associate degree are summarized as follows:

Electrical theory credited based upon M Test	= 32
Minimum Empire State College credits	= 24
Elective or transfer credits	= 08
Total required for Associate degree	= 64

This program is modeled after the IBEW's national program model.

U.S. Naval Facility Programs

Tidewater Community College and Thomas Nelson Community Colleges (VA) both offer programs designed to support the U.S. Navy's Naval Aviation Depot (see Textbox 8.1). Both colleges have designed their programs to meet the apprenticeship personnel training requirements of the Naval Depot. For instance, Tidewater Community College's offerings look like this:

Painter Curriculum		
One Semester	MTH	determined by test
Portsmouth Campus	ENG	determined by test
	CHM 195/01	Chemistry
	DRF 151	Drafting
	IND 115	Materials & Processes of Industry
	STD 100	Orientation
All Other Trades		
First Semester	MTH	determined by test (to qualify for avionics trade you must complete MTH 115 successfully)

	ENG	determined by test
	IND 115	Materials & Processes of Industry
	CIS 150	Fundamentals of Computer Info Systems
	DRF 151	Drafting
	STD 100	Orientation
Second Semester	MTH	Math
	ENG	English
	PHY 100	Physics
	DRF 152	Drafting
	MEC 120	Principles of Machine Technology

As can be seen here, much coordination is necessary between the educational institution and employer to ensure that testing can be done on each apprentice for proper placement.

COMMUNITY COLLEGE–INITIATED PROGRAMS

Community colleges are developing an increasing number of degree programs specifically for apprentices (see Textbox 8.2). These programs are tailored to incorporate the trade theory and on-the-job aspects of the apprenticeship into the degree requirements. For instance, at Sacramento City College in California, the following is offered:

> Sacramento City College offers a limited number of apprenticeship courses. All courses are applicable toward an Associate Degree. Enrollment in Apprenticeship courses is limited to those identified registered apprentices.
>
> Computer Science Apprenticeship Program
> The Computer Programmer apprenticeship program is offered in conjunction with the State of California Computer Programming Apprenticeship Consortium.
>
> *Teleprocessing and Data Characteristics* *3 Units*
> Prerequisites: Computer Information Science 2 or equivalent. Three Hours Lecture.
> Intended for employed data processing personnel who wish to receive updated training. Topics include: Business Data Communication Systems, functions and responsibilities of the data administrator, terminals on on-line and remote computing systems, data base concepts, file organization.
>
> *Data Processing Technical Work* *4 Units*
> Prerequisites: Computer Information Science 2 or equivalent. Three Hours Lecture.
> This course will familiarize the student with various technical tools which are

used in programming and the data processing environment. The course material will consist of lectures and hands-on experience with CICS, TSO Data Management command and PNVALET, PAN/CICS and TSO Text Editor Commands.

Job Control Language Utilities *4 Units*
Prerequisites: Computer Information Science 2 or equivalent. Three Hours Lecture.
This course will teach Job Control Language statements, how these statements relate to the operating system, and how to create JCL streams to execute programs. The course will also familiarize the student with various types of utility programs used in the data processing environment. (1992–1993 College Catalog, pp. 168–169)

For a student at Sacramento City College seeking a degree in computer information science, the degree requirements are sixty units. The student will then apply the units earned through the apprenticeship to the computer information science program and then proceed to earn the difference.

TEXTBOX 8.2

Wisconsin Technical Colleges and Apprenticeship

The sixteen Wisconsin technical colleges are very much involved in partnerships with the Wisconsin Department of Employment and Training and their local businesses for apprenticeship training. Each college offers all of the intake and processing services for apprenticeship.

The apprenticeship supervisor in the college works with the candidate apprentice to identify a participating employer, process the indenture agreement for the state, and enroll the apprentice in the appropriate related trades theory classes. A regular offering of trades theory courses are offered by each college. In addition, the apprentice can then continue on to earn an Associate degree if he/she so chooses.

Wisconsin has been a national leader in apprenticeship and has taken steps to move apprenticeship and dual-enrollment for Associate degrees closer together to provide a seamless educational path for the 21st century technician.

EQUATING CLASSROOM-RELATED TRAINING TO COLLEGE CREDIT

Montgomery College suggests the following guidelines for course and program development in a specific occupational area. First, identify a "home" for the program area within your college. It is always better for a program area to fit comfortably within the educational institution so that there exists some advocate(s) for the instructional subject area.

Next, analyze the subject area as previously described. Look for exit competencies that the apprentice must know and/or do upon completion of the courses and program. Now, compare what the apprentice does on the job to the exit competencies. Where there exists a parallel between coursework and job performance, room exists for academic credit recognition.

ARTICULATING CREDIT ARRANGEMENTS

Internal articulation agreements between apprenticeship programs and Associate degree programs are becoming more popular as a means to establish the college credit value of related education and training. Articulation involves a meeting of goals and objectives of both apprenticeship sponsors and college faculty. As discussed in Chapter 3, articulation involves communication. A result of effective articulation is a schedule of credit availability for related classroom education for the apprentice. Montgomery College grants up to twenty-eight credits in selected degree programs upon completion of the degree or certificate requirements and the full apprenticeship. The college uses the apprenticeship competency exam as evidence of successful mastery of the knowledge and skills associated with the courses for which credit is granted. Additionally, the student must remain in the apprenticeship as a craftsperson—not a supervisor—during the term of apprenticeship. Figures 8.7 and 8.8 (pp. 198 and 199) are sample letters to apprentice and program sponsor detailing such opportunities.

CREDIT FOR PRIOR LEARNING

Community colleges have also ventured forward to offer apprentices increased opportunities to earn credits towards an Associate degree concomitant with their experiences in the trade or occupation. Community colleges have done this in a number of ways. Following is a look at these various procedures for recognition of related education and training towards a degree.

ASSESSMENTS OF PRIOR LEARNING

Daytona Beach Community College (DBCC) uses the Assessment of Prior Learning Experiences (APLE) to help students earn college credit for skills and knowledge gained outside of a normal collegiate setting. Such work experience might have come from current or prior work experiences, military training, or other documentable methods or experiences. Vocational school experience may qualify for conversion to college credit under this process as well. Hence, apprentices can apply previous classroom clock

Montgomery College

December 5, 1994

Mr.
Local Steamfitters Joint Apprenticeship Committee

Dear Mr.

After careful review of the comprehensive, Steamfitters Apprenticeship Program, we are extremely impressed with the program design, content and structure. We are pleased to confirm that your program of study exceeds a comparable level of skills and knowledge equivalent to a series of related college credit courses at Montgomery College.

These courses are:

		Credits
CT130	Construction Methods and Materials	3
CT131	Construction Plan Reading	3
CT283	Mechanical & Electrical Systems	3
BU130	Introduction to the Building Trades	3
BU144	Electricity I	4
BU146	Plumbing I	4
BU244	Electricity II	4
BU246	Plumbing II	4
BU250	Safety for the Building Trades	3
Total Equivalent Credits:		31

The Steamfitter Apprentice Graduate who presents formal certification of completion of your program will be granted these credits upon completion of their course of study at Montgomery College.

Thank you for assisting in the creation of this critical industry-education linkage.

cc: Director of Admissions & Registrations

Sincerely,

The Gudelsky Institute for
Technical Education

Central Administration 900 Hungerford Drive Rockville, MD 20850 (301) 279-5000 | **Germantown Campus** 20200 Observation Drive Germantown, MD 20876 (301) 353-7700 | **Rockville Campus** 51 Mannakee Street Rockville, MD 20850 (301) 279-5000 | **Takoma Park Campus** 7600 Takoma Avenue Takoma Park, MD 20912 (301) 650-1300 | **Continuing Education** 51 Mannakee Street Rockville, MD 20850 (301) 279-5188

Figure 8.7 Articulation letter. (Reprinted with permission from George M. Payne, Montgomery College.)

Montgomery College

Dear Apprenticeship Program Graduate

Congratulations on your program completion. This represents an outstanding commitment that was made by you, your employer, and your family. By now you fully recognize the need and value of continuing your education to remain current and to increase your capabilities in your career field.

The program you have just completed has established an articulation agreement with Montgomery College to provide an opportunity for you to receive college level credit for the technical training content areas you have now mastered. A copy of the articulation agreement is attached.

The individual courses for which you can receive credit from this articulation agreement are partial program requirements for Certificate programs, typically 24-32 credit hours, or for Associate of Applied Science degree programs, typically 60-68 credit hours. The remaining degree requirements include both technical and general education coursework.

The primary programs which the identified courses apply toward include:
- *Building Trades Technology Certificate*
- *Building Trades Technology Associate of Applied Science*
- *Management of Construction Certificate*
- *Management of Construction Associate of Applied Science*

When you are ready to apply your apprenticeship training toward a college degree here are the steps you should take:
1. *Review the current College catalog for degree requirements.*
2. *Meet with the appropriate program coordinator.*
3. *Confirm which courses from the attached agreement apply to the program of your interest.*
4. *Register for classes.*
5. *Apply for graduation.*
6. *The articulation credits are awarded upon completion of your Certificate or AAS degree*

For additional information, please call Mario Parcan at 301-279-5142 for Management of Construction or Ed Roberts at 301-251-7688 for Building Trades Technology.

Again, congratulations on your achievement and never stop learning!

Central Administration	**Germantown Campus**	**Rockville Campus**	**Takoma Park Campus**	**Continuing Education**
900 Hungerford Drive	20200 Observation Drive	51 Mannakee Street	7600 Takoma Avenue	51 Mannakee Street
Rockville, MD 20850	Germantown, MD 20876	Rockville, MD 20850	Takoma Park, MD 20912	Rockville, MD 20850
(301) 279-5000	(301) 353-7700	(301) 279-5000	(301) 650-1300	(301) 279-5188

Figure 8.8 Letter to students. (Reprinted with permission from George M. Payne, Montgomery College.)

hours towards a degree through this process. DBCC states in their 1994–1995 catalog:

> Students who want to apply for APLE must be currently enrolled at DBCC or have earned 12 credit hours at DBCC, excluding college preparatory and vocational preparatory credits. Credits may be applied toward general education, degree or certificate program requirements. Degree-seeking students may earn a maximum of 40 credit hours through APLE. A maximum of 25 percent of the credit hours required for a degree may be earned through portfolio based credits. Additional portfolio credits may be considered in exceptional cases. A maximum of 60 percent of the hours required to complete a certificate program may be earned through APLE. These credits may not be used to meet the graduation requirements of 25 percent of credit hours in residence.
>
> For A.A. and A.S. programs, students must have completed at least 12 credit hours at DBCC before credit will be posted. For certificate programs which require fewer than 20 total credit hours, students must have completed at least 12 semester hours before credit will be posted. (p. 22)

COLLEGE-LEVEL EXAMINATION PROGRAM (CLEP)

Apprentices may earn college credit by scoring well on CLEP tests. CLEP testing is designed to allow students who have a particular subject matter expertise to challenge the course by passing an examination covering the particular course content.

COOPERATIVE EDUCATION

Structured work-based education, oftentimes termed cooperative education, is another way in which students are able to work and learn. In some colleges, cooperative education and apprenticeship are almost synonymous. In the automotive technician arena, many motor companies and dealerships engage apprentices in cooperation with the community college through cooperative education. As previously described, Toyota Motor Corp., Ford Motor Company, American Honda Motor Corp., Chrysler Corp., and General Motors have such programs operating cooperatively with vocational-technical schools and community colleges.

INDEPENDENT STUDY

In some cases, some colleges will permit apprentices to gain some college credit in a specific area through independent study. Here, the student prepares a paper or some other form of project working essentially independently, though under minimal supervision.

Many community colleges now also directly award degree credit for "hands-on" portions of apprenticeship training. At some colleges such as American River College, the related instruction is listed in the college catalog with equated college credit. American River College (CA) provides the following:

> American River College offers a number of apprenticeship related training programs. Given below are the courses required for receiving the Certificate of Achievement in each of these programs. Completion of graduate requirements (listed on page 38) in addition to the Certificate of Achievement requirements, will qualify you for an A.A. degree. (1992–1993 Catalog, p. 87)

Long Beach City College (LBCC) supports the U.S. Navy's Shipyard trades skills apprenticeship programs. LBCC offers the following course to equate shipyard hands-on work towards an Associate degree in ship construction and repair.

> *Supervised Shop Experience (2) Four Years*
> 150 Hours work
> Prerequisite Open only to LBNSY Apprentice Program enrollees.
> Supervised shop experience in one of 20 apprentice training programs (see list) granting 2 units of credit per semester, maximum of 16. This course is an occupational preparation experience designed to meet specific unit requirements and is not conducted as a transfer level course. (1984–1985 Catalog)

CLASSROOM LOGISTICS

Employers want apprentices' related education and training to be as easy to administer as possible with ease in terms of scheduling and in terms of locations. Therefore, it is incumbent upon the educational partner to plan instructional delivery to be as user friendly as possible for apprentices. Considerations include:

- work schedules of apprentices
- numbers of apprentices requiring related education in each subject area or module or course
- distance of work site(s) to educational institution's classroom
- availability of classroom space in employer(s) facilities
- availability of required laboratories in either educational institution and/or employer facility
- ability to commingle apprentices from several programs to plan and schedule related education efficiently

Consider each of these variables when making decisions on how to schedule related education offerings.

SCHEDULE FOR CLASSES

While some apprenticeship programs and educational institutions reserve classroom activity for the evening hours, others prefer to make the classroom learning an activity on par with the on-the-job training and provide both activities during the day. For instance, Northcentral Technical College in Wisconsin offers:

> As an apprentice, you will be supervised on the job by skilled journeypersons. You will also attend day school for four hours each week or eight hours every other week. You will receive your normal hourly salary while you attend class during the day. Some apprenticeship programs also require you to attend night school which you will have to pay for yourself. Apprentices usually attend 300–600 hours of classroom instruction during the length of their apprenticeship. (1990–1992 Catalog, p. 73)

AVAILABILITY OF FACILITIES

Planning also involves facility coordination and maintenance. Instructors will be sharing facilities and therefore need information about communicating needs for supplies, repairs, and other like concerns. Montgomery College, much like other educational institutions plans for this through the use of an Instructor Handbook. The handbook contains information and forms regarding:

- lab maintenance request form
- safety/lab/equipment procedures
- standard operating procedures for the course

CLASS BEHAVIOR

Instructors should be aware of institutional procedures for dealing with student problems. Sometimes, it becomes necessary to counsel apprentices about the appropriate behavior in classes. A statement that appears in the Empire State College–IBEW Local #3 Apprentice Handbook that is useful follows:

> If a faculty person is dissatisfied with the conduct of a student in class, the faculty person has the right to eject that student from class and to send the student to the administration office. . . . Depending upon the result of that counseling, a student may be allowed to return to that class. Other possibilities—again, depending upon the nature of the problem—include being assigned to another class or being administratively withdrawn from the school.

STUDENT GRIEVANCE PROCEDURES

Students also have rights, and a procedure for dealing with apprentice student grievances also needs to be considered in program development. Therefore, consider these steps for incorporation into your student handbook and standard procedures for your JAC or educational foundation program. Of course, student grievance procedures are undoubtedly already in place in Voc-Techs and community colleges, and if you are operating under the aegis of such an educational institution, then you will use those procedures. These include:

(1) Informally discuss the problem with the instructor(s) or work-site supervisors involved. If the matter is not resolved, then follow the next step.
(2) Discuss the matter with an educational counselor. The counselor will bring both parties together and attempt to resolve the matter. If this doesn't work, then go to the next step.
(3) Bring the matter to the school/college administration. The administration will again bring both instructors and student together to resolve the issue. If not successful, then the next step follows.
(4) Bring the matter to the appropriate academic review committee. The committee usually has some legal authority and will proceed according to a stated policy. Alternatively, the student can follow step 5.
(5) Bring the matter to the academic vice president. This individual will exercise institutional authority to resolve the problem. This person's decision in the matter is usually final—much like the academic review committee.

GRADUATION

Graduation from apprenticeship programs deserves recognition. Voc-Techs and community colleges also recognize the value of apprenticeship program support by offering graduation ceremonies for graduating apprenticeship classes. Montgomery College regularly offers a formal ceremony wherein all of the participating employers are invited to celebrate their apprentices' achievements. Certificates are presented at the ceremony (see Figure 8.9, p. 204). Oftentimes, dignitaries such as the state DOL representative attend and speak.

TRAINING WORK-SITE SUPERVISORS AS INSTRUCTORS

A very important service that the educational partner can provide to the cooperative apprenticeship partnership is training of work-site supervisors as effective instructors. Some states provide funding for apprentice instructor development. In many cases, these funds are targeted for the participating educational institution—the community college or teacher training institu-

Montgomery Community College

Office of Continuing Education

APPRENTICESHIP RELATED INSTRUCTION PROGRAM

This is to certify and present to

on this day of

"CERTIFICATE OF COMPLETION OF RELATED INSTRUCTION"
for the successful achievement of the required technical and theoretical subjects for the trade of

This instruction has been completed in accordance with the Standards of Apprenticeship and Training, which has been approved and registered with the United States Department of Labor Bureau of Apprenticeship and Training and/or the Maryland Apprenticeship and Training Council.

Figure 8.9 Certificate of graduation. (Reprinted with permission from George M. Payne, Montgomery College.)

tion. Certain professions also provide training assistance. For instance, in the firefighter professions, instructor development is recognized as essential. California has a series of instructor development courses that cover the basics of instructor training and lead to certification as a firefighter instructor. These courses also enable these individuals to be certified as community college instructors. Unions such as the United Association offer formal training programs and certification for their apprentice instructors. Union national offices should be contacted to determine the availability of such programs.

The degree to which instructor development is useful and productive really rests with the numbers of apprentices that a particular supervisor will eventually train. While this might be difficult to predict, you might consider phasing in a program of supervisor training, thus providing more successive training as the supervisor becomes more involved with apprenticeship activities.

What are some of the areas of expertise that the supervisor should be training in?

(1) Roles and responsibilities of the effective instructor: Work-site supervisors are the most important element in the apprenticeship program. As such, they should understand their importance and the role that they play in training delivery. The supervisor should understand the critical responsibility charged to him/her. Instructors should be familiar with procedures for: student attendance records, grades, instructor absences, school closing, course evaluations, syllabus requirements, doing activities/exercises, and student participation.

(2) Effective communications skills: The work-site supervisor must get the message across. Effective spoken and written skills are a must. Of course, you are in no position to suggest that an employer or employee supervisor needs work in this area. You can, however, provide short seminars or courses available for participating employers at no cost, as part of the quid pro quo.

(3) Effective skills demonstration techniques: The processes that an instructor uses to demonstrate a skill can be reasonably and easily taught to any craftsperson. We advocate the five-step skills demonstration technique (Cantor, 1992):
   - The instructor demonstrates the particular skill in real-time, as it is actually done on the job—not slowly and without any commentary. It is done as it is usually done.
   - Next, the instructor demonstrates the skill slowly and explains what is being done to the learner.
   - Then, thirdly, the instructor does the skill and the apprentice states the step in the skill after the instructor has done it, for verbal reinforcement. Here, the instructor does, and then the apprentice says.
   - Then, in step 4, the instructor allows the student to call out the steps before he does the skill. Here, the apprentice says, and then the instructor does.
   - Then the apprentice practices the skill under the instructor's eye to perfect the skill. The instructor intervenes only if and when necessary. Figure 8.10, on p. 206, lists these steps.

(4) Psychology of the adult learner

(5) Evaluation of apprentices

Again, you can provide short courses or seminars to participating employers in adult psychology. You can also integrate some of this information into

DEMONSTRATION
METHODOLOGY FOR TEACHING A
PSYCHOMOTOR SKILL

- INSTRUCTOR DEMONSTRATES SKILL IN REAL TIME
- INSTRUCTOR DEMONSTRATES SKILL IN STEPS
- INSTRUCTOR DOES - STUDENT TELLS
- STUDENT TELLS - INSTRUCTOR DOES
- STUDENT PRACTICES SKILL

Figure 8.10 Demonstrating a skill.

the debriefings you will be doing with employers and supervisors as part of the ongoing monitoring of apprentices that you will be doing. Where feasible, make suggestions about dealing with apprentices to supervisors, which will make it easier for them to understand human interactions for training.

- Use learning objectives.
- Use and develop lesson plans.
- Design and manage learning activities.

The three areas of training can only be effectively done in a short course or seminar. It is suggested that, for those on-the-job supervisors who will be participating in the program for an extended period of time, a no-cost seminar can be acquired or designed to provide them with these instructional skills. Again, as said earlier, such programs do exist in some industries for on-the-job supervisors, such as the fire service and nuclear power industries. These two industries recognize training as essential to human performance, and they invest time in training their trainers.

Apprentice instructors can also gain useful skills through in-service training courses offered at local universities. Consult the local teacher's college or university in your area for information on the types of courses available for adult/vocational educators.

SUMMARY

This chapter has described the essential elements of related education and support for the apprentice. The chapter has discussed both traditional related classroom instruction and dual-enrollments of apprentices for college credit and Associate degrees.

REFERENCES

California Firefighter Joint Apprenticeship Committee. (1989). *Program Manual*, Sacramento, CA: Author.

Cantor, J. A. (1992). *Delivering Instruction to Adult Learners*. Toronto, Canada: Wall & Emerson, Inc.

Central Piedmont Community College. (1991). Toyota Technical Education Network Program: 1991 Dealer-Student Information Package. Charlotte, NC.

Empire State College, SUNY. (Undated). *Apprentice Handbook*. New York, NY: Author.

Massachusetts Bay Community College. (1992). Massachusetts Bay Community College/General Motors Corporation/Minuteman Regional Vocational Technical School General Motors Automotive Services Educational Program Student Information Packet. Boston, MA.

MechTech, Inc. (1995). Program Reference Materials. Cranston, RI.

Montgomery College. (1987). Apprentice Program Descriptive Materials and Brochures. Rockville, MD.

Pooler, D. (1993, Aug./Sept.). "On Target: The Apprenticeship Approach," *Technical & Skills Training*, pp. 29–30.

Sacramento City College. (1992–1993). College Catalog. Sacramento, CA.

U.S. Navy, Naval Aviation Depot. (1990). *Cooperative Apprentice Training Program: Handbook*. Norfolk, VA: Author.

U.S Department of Labor, Manpower Administration. (1975). *Dual Enrollment as an Operating Engineer Apprentice and an Associate Degree Candidate*. Final Report. International Union of Operating Engineers, Washington, D.C.: Author.

Apprentice Testing and Evaluation and Certification

ANOTHER key aspect of successful cooperative apprenticeship is testing and evaluation of the apprentice on the job. Evaluation, in turn, leads to certification. The design of appropriate evaluation instruments will be presented in this chapter. The information in this chapter is as follows:

- The basics of testing and evaluation—valid and reliable measures of an apprentice's performance based upon an occupational work schedule.
- Testing related education—written tests of achievement in classroom-related work.
- Testing on-the-job performance through performance tests. Development of competency checklists to complement performance objectives for on-the-job training use will be presented. Supervisory use of these evaluative tools on the job will be highlighted.
- Related issues of certification of apprentice competency including liability issues—especially for youth apprenticeship. Certification of apprentices will be included.

THE BASICS OF TESTING AND EVALUATION

Testing is a necessary component of cooperative apprenticeship programs. The writing of test items is a necessary skill to ensure that the test is both content valid and a reliable measure. We evaluate apprentices for several reasons.

Testing and evaluation is done at three specific points in an apprentice's education and training: at entrance into the program for guidance and counseling purposes; periodically during on-the-job training and related classroom portions of the program for apprentice feedback, diagnostic, and progress measuring; and at the end of the program for evaluation and certification purposes to measure job readiness. This permits us to certify the apprentice as a journeyman ready to formally enter the trade or occupation.

TWO ESSENTIAL ELEMENTS

The process of designing and developing tests—whether for assessment of related education (knowledge) or on-the-job (motor skills)—involves the writing of multiple items (questions). Most jobs and tasks demand knowledge and competence in multiple skills and information areas; therefore, you will ultimately assemble numbers of items that indicate total mastery of a subject area. This will ensure a test's validity and reliability.

Validity refers to the test instrument's content; in other words, the test should measure what it claims to measure. The test should assess the knowledge, skills, and abilities required to function competently on the job. Any effort on your part to measure more or less than is required to do the job diminishes the validity of the test. To ensure test validity, we are going to enlist the aid and assistance of subject matter experts to design and develop the test. These experts are going to be job incumbents—journeymen and supervisors of journeymen representing a cross section of the trade for which the test is being developed.

As validity determines whether or not a test measures what it is designed to measure, reliability determines the degree to which it does so in a consistent manner. A test like any other measuring instrument is useless unless it can produce results consistently. To ensure test reliability, we are going to make suggestions about how to properly administer the tests.

TESTING RELATED EDUCATION VIA WRITTEN TESTS

Regardless of the purpose of the test, development of tests also proceeds through a three-phase test development process (Cantor, 1987). To ensure test validity, you should commence the process by calling a meeting of people representative of the occupational trade area. While the process that is described herein is very comprehensive and best suited for the final certification exam, you will gain information useful for classroom or intermediate skills tests as well.

THE TEST DEVELOPMENT TEAM

These people must be knowledgeable of the area and should be at least journeymen. Some supervisors should also be part of the team. Make sure that the people on this committee are representative of all aspects of the trade or occupation and are themselves well-versed and competent in their trade areas.

In this first phase, you will gather all of the information necessary to build the test. You will identify:

- the skills performances or related information to be tested
- the conditions under which these skills or knowledge should be demonstrated (and therefore tested)
- the degrees or standards of required performance

This comes from the task analysis you performed earlier (see Chapter 5). It must parallel the work processes schedule for the occupation for which you are building the test.

PREPARE A TEST DESIGN SPECIFICATION

Integral in this phase of test development is development of a test design specification. I suggest that a test design specification listing the requirements for the whole test be developed. A test design specification identifies what is to be tested. This document will aid you in providing backup evidence that the apprentice's certification is, in fact, based upon sound criteria drawn from the trade and occupation. To assemble such a document, each test item should also have a specification detailing the behavior, conditions, and standards for apprentice performance. In drafting the test design specification, together with the committee, you will identify:

- the purpose of the test (e.g., apprenticeship quarterly evaluation, related education unit, or final mastery test)
- the specific behaviors (jobs/skills or information) to be tested (from the task analysis data and from the written performance objectives per Chapter 5)
- conditions under which the behaviors are to be performed (e.g., tools/materials)
- standards to which the behaviors must be performed (e.g., $\pm .003$, or eight out of ten items correct from industry documentation)

The test design specification will also specify which behaviors will be tested by written test and which will be tested by performance test. A test design specification provides the evidence that the evaluation is based upon "best industry standards and practices" and is fair and impartial.

It is useful to begin the design process by reviewing the work processes schedule and job/task analysis. In some instances you might wish to then divide it up among the committee members based upon the committee's membership and expertise. Perhaps you could put each major job on an index card and then spread the cards out on the table. Consider each job in terms of:

- criticality of the job and its component tasks
- difficulty of the job and its component tasks
- frequency of performance of the job and its component tasks

You might wish to rate each of these points independently for each job on a 1–5 scale.

Criticality of Performance

If a job is critical to the performance of a trade or occupation, then you would probably want to ensure that the apprentice can do it properly prior to exiting the training program as a certified journeyman. Hence, you will want to develop test items for that job.

Difficulty of Performance

How difficult is a job/task to master and perform? If it is deemed a difficult job, then does the committee feel it important to ensure that the apprentice can perform it properly and, thus, devote a number of items to that job? Conversely, if it is an easy task and the apprentice really does not perform it too well, he/she might be able to master it easily when finally on the job as a journeyman. Therefore, less testing might be indicated.

Frequency of Performance

How frequently is a job performed? Jobs that are performed often might require intense testing if they are critical to performance of the occupation and/or if they are difficult to master and perform. Conversely, an easy-to-do job, performed infrequently, might not require much testing attention.

Now score the findings of this exercise. Decide upon the jobs/tasks and information to be included on the test and the number of items to comprise the test. Rule of thumb: about sixty items can be answered in a written test in one hour.

Conditions of Performance

For each job/task and informational item, have the committee provide insights into the background conditions of performance. Ask them the following:

- What tools, materials, test equipment, physical surroundings,

formulae, books, documents, weather conditions, and so on should be present for the apprentice to respond to this item?
- Must this job/task or informational item be performed outside or inside or as part of an evolution, scenario, or reenactment?
- Is a written test or performance test indicated here?
- Is a laboratory required for testing purposes?

Record the findings on the index card.

Degree or Standard of Performance

Standards are often a difficult piece of information for a committee to provide. What you are actually asking is: What is an acceptable standard of performance for this particular job when the apprentice is actually on the job as a journeyman? What will industry accept as best industry practice for this job? The answer might be specified in terms of:

- time to perform the job and associated tasks
- numbers of units of something produced
- measurements or tolerances to a blueprint
- numbers of acceptable errors in a specified period of time or unit produced
- published standard

In any event, record the committee's response and the sources of authority that they refer to for the particular responses. You will need these if the test is ever challenged by an apprentice or another interested party.

TESTING RELATED EDUCATION INFORMATION

DRAFTING TEST ITEMS

The next phase is the actual drafting of the test items. You can reserve the writing of test items for yourself, or you can use the committee for the actual writing. Usually, a committee of tradespeople will not wish to sit and write test items. Therefore, it is best that you solicit the needed information for the item writing from the committee and then put someone else, well-versed in writing, to work on the actual task of writing the test items.

Again, item writing is an art (Cantor, 1987; 1988; 1992). While no single set of rules can ensure good test items, some general principles can help. Your skill in applying these principles determines the quality of each item and, as a result, the integrity of the test as a useful measurement instrument.

In this phase, you might be developing related information tests in which the apprentice will respond in writing and tests of physical and/or motor skills in the form of checklists to be performed by the apprentice. Following is a look at developing written test items.

MULTIPLE-CHOICE TEST ITEMS

By far the most popular format for written tests because it can determine mastery of many different subject matter areas and assess a variety of cognitive processes from straightforward factual knowledge to complex thinking processes is the multiple-choice test (Cantor, 1987; 1992). Multiple-choice tests comprise a series of questions called test items. Each measures a particular piece of information, providing specific guidance about what a person knows about the topic. To be valid and accurate, a good test must rely on well-written items.

Format of Multiple-Choice Items

A multiple-choice item consists of two parts: (1) a stem and (2) a set of distractors or responses, one of which is correct. The sample below represents a typical multiple-choice item.

The stem has a single central theme, in this case the name of the first president of the United States. It asks a question, presents a problem, or takes the form of an incomplete statement.

Who was the first president of the United States? (stem)

A. James Madison	(distractor)
B. John Adams	(distractor)
C. George Washington	(correct answer)
D. Abraham Lincoln	(distractor)

Test developers prefer the question form because it presents the central theme clearly, and it avoids tipping off or cuing the correct answer. The distractors list possible answers to the problem or question. Here, they provide a variety of choices of the first president of the United States. Notice that the correct answer and the three distractors are similar, both in terms of sense and grammatical structure. This points out a crucial point: all distractors must be plausible as correct responses, and they must be written in a way that prevents apprentices from guessing the correct answer simply on the basis of the distractors' format.

Knowing this basic format—a stem with one central theme, one correct answer, and three similar and plausible distractors—can help get you started on good test item construction, but first, you have to decide what kinds of information the test will cover.

Central Theme and Possible Distractors

Together with your test writing committee, you need to identify the central themes to be tested. Refer back to your test design specification at this point. Also be sure you have all the necessary technical references, including the work schedule, at hand. Use these to identify the aspects, concepts, and facts that are critical to the mastery of the subject matter area. These critical skills or knowledge form the core of your test; each of them represents a possible central theme for a test item.

Now, ask yourself some questions:

- How can an apprentice demonstrate knowledge of this theme?
- In what sort of circumstances might it be important to understand these concepts?
- What might be the consequence of a journeyman not knowing or understanding these concepts?
- What do employers say are some of the common misconceptions or misunderstandings about this subject?

These questions and their answers help you to fine-tune the test items.

For example, suppose the subject area is firefighting and your test design centers on the area of positive pressure ventilation. You will ask yourself, "If someone did not know the proper way to establish positive pressure ventilation, what might he or she do?" Write down the answers that occur to you and then all the correct responses to your test. This information becomes part of your test item specifications. This now makes a good preliminary set of distractors.

At this point, don't be overly concerned about proper wording or format: you are only trying to create a first draft of the items, like the one shown below:

A structure is fully involved in fire. The command has been given to establish positive pressure ventilation. What is the first step that should be done?

A. Break out all of the first floor windows. (distractor)
B. Determine the desirable airflow pattern. (correct answer)
C. Break through the roof. (distractor)
D. Break out all of the top floor windows. (distractor)

Write out the stem, making sure it presents a single central theme. Although the sample asks its question after two separate sentences, it has one central theme: What should you do to create positive pressure ventilation? Now, arrange all your distractors, including the correct response, below the stem and letter them consecutively. If more than the four choices needed occur to you, put them down. You can decide later which are the best. The test item in the sample needs revision, and your first draft will too. But now you have a base from which to construct your final version of the item.

Item Format and Stem Revision

Multiple-choice items commonly take four forms: correct answer, best answer, negative, and combined-response. The type of item you select will depend on the material to be tested and the complexity of the process that indicates an apprentice's mastery. In general, though, the simpler and clearer the item is, the better.

No matter what format you choose for your test items and their stems, keep a few general rules in mind. Include only one central theme in each stem. This lets the apprentices zero in on the exact information you seek. Word each stem as clearly and concisely as possible to help avoid confusion. If there are any words or phrases common to all distractors, place them in the stem instead. Also include in the stem any qualifying words that limit or isolate the possible responses.

Developing and Writing Distractors

Now you have written a good stem, formulated a correct response, and collected a preliminary list of possible distractors. Review that list again, remembering that all distractors must be plausible and must fit the grammatical structure of the stem. Some general guidelines can help you revise your choices.

First of all, avoid using "none of the above" or "all of the above" as distractors. They are often confusing. Also avoid obviously incorrect responses. Both of these devices destroy plausibility.

To make sure your distractors are plausible, define the class of things to which all of the answer choices should belong. For example, go back to the second sample, the item that asks you what you should do to ventilate a fully involved structure. The class of possible answers might be defined as "ventilation measures." This process gives you distractors A, B, and D, which seem like valid ventilation procedures.

Now, think of things more specifically associated with the terms in the stem and consider some misconceptions people may hold about the subject in question. Now, you've got distractors A and D because an uninformed person might think the most important factor is letting heat out of the structure. Thus, all the distractors are plausible. Most test items contain only three distractors plus the correct response.

Remember to write distractors so they make grammatical sense. That means making them correspond to the verbal structure of the stem. Be sure that the beginning of each distractor logically follows the last word in the stem. For example, if the stem ends with the word *an,* apprentices will probably assume that the correct response is a singular noun beginning with a vowel; make sure all the distractors have this pattern. When using items that ask apprentices to complete a sentence, ensure that all the distractors do, in fact, complete the sentence. Also note in the second sample that all of the distractors begin with capital letters and end with a period. Finally, place all distractors in logical order if they are numerically or sequentially related.

Remember that, like stems, the most effective distractors are clearly and concisely written, without excess information that could confuse the apprentice. Keep in mind that a good test item measures the apprentice's ability to recall a central theme, not his/her ability to take tests.

MATCHING TEST ITEMS

Matching test items can condense the testing of a relatively large amount of information. These tests require information sequencing and assimilating. When the facts are selected carefully, the chance of guessing is greatly reduced. Matching test items cannot effectively test higher order intellectual skills, and their content must have relational association (i.e., cause and effect, term and definition).

Matching test items (see sample below) are related to multiple-choice test items in that many of the latter can be combined into one matching test item. They are good for testing content that includes classifying, sequencing, labeling, or defining (i.e., defined concepts). Although matching items are an acceptable method of assessment, avoid their overuse.

Directions: Match the type of test on the left with the disadvantages of using the method on the right.

____	1. Multiple choice	a. Can't test higher levels of learning.
____	2. True/False	b. Takes time to construct effective test.
____	3. Matching	c. Questions can be too broad or ambiguous.
____	4. Short answer	d. Can only test factual information.
____	5. Essay	e. Questions must be carefully phrased to lead apprentice to a specific answer.

A well-constructed matching test item contains clear instructions stating how the items are to be matched and whether responses are to be used only once. Each question should have a central theme with parallel problems and parallel responses: this parallelism will help prevent guessing. If used, distractors should be plausible. Make sure there are no grammatical clues (e.g., "a" or "an") given in the items. Arrange alternatives in logical order, for instance, alphabetically or numerically. This ordering makes it easier to locate alternatives and reduces the chance of giving subtle clues.

SHORT ANSWER/COMPLETION ITEMS

Short answer/completion test items allow a large amount of information to be condensed into a few questions. Because the apprentice must write something, you can make a clear assessment of what has been mastered; however, these items are limited to recall of specific facts. The items must be carefully phrased to avoid directing the apprentice to a specific answer.

List the steps performed in connecting an ohmmeter to a circuit.

1. ______________________________
2. ______________________________
3. ______________________________
4. ______________________________

Short answer/completion test items are easy to construct and may be used to test recall of specific facts, concepts, or principles (see sample above). When constructing completion items, be certain that there is only one answer that will complete the statement. Answers should require no more than a few words; if more are needed, it will be hard to grade the answers objectively.

A common structure for the short answer/completion test item is an incomplete statement. In this form, a significant word or two is left out, and the apprentice must complete the statement. Each test item should be matched to a performance objective.

ESSAY TEST ITEMS

Essay test items (see sample) can test for higher order intellectual skills and cognitive strategies. They allow your apprentices to display mastery of an entire range of knowledge about a subject. There are several disadvantages also associated with essay tests. First, it is often difficult to write question items that are not too broad or ambiguous; this in turn makes scoring difficult. Second, only a limited number of concepts and principles can be tested. Essay tests are not fair to apprentices with poor writing skills. From an instructor's point of view, essays are very time-consuming to assess.

- Discuss the four primary advantages of an incident command system and compare these advantages to the alternatives of not using an incident command system. (Each advantage and alternative discussed will be worth 4 points.)

While essay test items are excellent for evaluating higher levels of learning and application of principles, they are very difficult to construct in such a way as to limit subjectivity. In a well-constructed essay test, complete directions are given, questions are clearly and unambiguously written, model responses are used in grading, and components and their weighing are identified. Ample space is also provided for the answer, or directions are given for using additional space. Allow apprentices plenty of time to formulate and construct their answers.

TESTING ON-THE-JOB PERFORMANCE THROUGH PERFORMANCE TESTS

SKILLS PERFORMANCE CHECKLISTS

Skills performance checklists are the assessment devices for identifying and recording data about apprentice skills performances in specific motor skills (job performance) areas. By first analyzing the purpose of the performance test (test specification) and the behaviors (specific performance objectives) to be measured, a determination of the best approaches for testing can be made. Two types of performance tests are commonly used—product or process. A product performance test requires the apprentice to actually produce an item or perform a series of tasks associated with the job from start to finish. A process performance test requires only a series of steps or tasks to be demonstrated.

Most performance tests are either of a pass/fail nature, in which performance measurement is based on an established criterion or standard, or of a normative nature in which the measurement of performance will compare each apprentice's skill mastery or competence with other apprentices.

The test design specification for a skills performance test must indicate the performance-based behaviors to be tested. To do this and be able to identify the motor skills, abilities, and associated knowledge critical to carrying out the work, it is necessary to analyze all pertinent instructional objectives in detail. This analysis is an important part of establishing the content validity of the test. The test design specification should identify:

- the specific behavioral elements that should be measured
- whether a performance test or written test or both is actually most appropriate for the measurement

Evaluation checklist

Name: ____

Trade: Machinist

Title: Slotted Block

Step	Checkpoints	Checkpoint Performance		Required Step Score	Maximum Possible Score	Actual Score	Step	
							Pass	Fail
A.	Set Up Vise	Totals		5.5	11			
	1. Mount Vise: Depth 1-7/8	Sat.	Unsat.		2			
	2. Indicate vise parallel to table slots a. Secure indicator (dial test) to machine quill using drill chuck. b. Snug nuts holding vise to table. c. Bring quill down to touch solid jaw of vise, putting about .005 pressure on indicator. Set indicator to 0. d. Rapid traverse table longitudinally. Read indicator. Tap vise to remove any differences in reading. Rapid traverse in opposite direction. When a zero reading on indicator is achieved, tighten nuts holding vise. e. Reindicate to ensure that vise has not moved while tightening.	Sat.	Unsat.		7			
	3. Scale block to be machined to determine if adequate material is available	Sat.	Unsat.		2			
B.	Set Block in Vise	Totals		4	8			
	1. Use parallels of same height (approximately 1") against solid movable chuck jaws	Sat.	Unsat.		2			
	2. Lay down block on parallels with 1" dimension up and about ½" extending beyond end of vise	Sat.	Unsat.		3			
	3. Tighten vise	Sat.	Unsat.		2			
	4. Tap block down on top	Sat.	Unsat.		1			
C.	Insert End Mill	Totals		3	6			
	1. Select 1" diameter end mill	Sat.	Unsat.		1			
	2. Measure cutter shank and install proper collet in quill	Sat.	Unsat.		2			
	3. Install cutter shank into collet and tighten drawbar	Sat.	Unsat.		1			
	4. Make sure quill is completely withdrawn	Sat.	Unsat.		1			
	5. Lock quill	Sat.	Unsat.		1			
	Evaluation Sheet Totals							

Figure 9.1 Assessment checklist. (Copyright © Jeffrey A. Cantor. From Cantor, J. A. [1988, September). "How to Design, Develop, and Use Performance Tests," *Training & Development Journal.*]

- the number of questions or trials needed to establish proficiency on a task

MEASUREMENT POINTS

Ask several qualified experts on your test design committee to review and agree upon the critical measurement points (based on the instructional objectives and task analysis data) to be used. Establish a process for recording relevant work and training experience of the people who will judge test content. This aids in a test's content validity.

Each expert should record which task analysis components are important to the overall purpose of the test and should be measured. These components include duties, tasks, task elements and knowledge, skills, and abilities. For example, to measure an apprentice's performance on a specific task, such as overhauling a computer motherboard, tasks and task elements probably would be rated as more important to the end product than to the analysis. If overall job performance is to be measured, then the knowledge, skills, and abilities would be rated higher. The level of detail in the analysis need only be deep enough to satisfy the intended purpose of the test.

THE CHECKLIST

Once the test design specification is completed, you will need to decide how much time will be required to compile test materials and processes to complete the test. Record all such information to ensure test validity. Testing materials might include a proctor's or supervisor's guide, directions, information sheets, drawings, tools, machines, and so on. It will also include a checklist to record apprentice performance. Write notes on the test item form or checklist as appropriate, such as technical standards, safety measures, constraints, and scoring methods. Figure 9.1 is a skills performance checklist.

REVIEWING THE ITEMS AND TEST

Now that you have fine-tuned your draft item, you need to review it. Ask yourself the following questions:

- Does the item measure what I am trying to measure?
- Does the stem contain just one central theme? If not, you may need to reword it or split it into more than one item.
- Do the apprentices have all the information they need to answer the item?

- Will the item's intent be clear to someone reading it for the first time?
- Is the wording as clear and concise as possible? If not, can the item be revised and still be understood?
- Are the distractors plausible, and do they make grammatical sense? Are any of the distractors clearly inappropriate?
- Is the correct answer clearly correct and up-to-date according to recognized experts in the field?

If your test item satisfies all these standards, it is a good one. Once a test is developed, it should then be pilot tested to ensure that it does, in fact, test in a consistent manner what it is supposed to test.

RELATED ISSUES OF CERTIFICATION OF COMPETENCY

A means of determination of an apprentice's competency is extremely important to the apprenticeship system. After all, the issuance of a certificate of completion means that the apprentice is now at journeyman level in the trade or occupation. The sponsor, employer, DOL, and probably the educational institution are certifying to that contention; therefore, the development of a valid and reliable test to assess such competence is essential. To meet this challenge, we have described processes for test development. Feedback from employers, in addition to the final exam, is also essential; therefore, Figure 9.2 illustrates a form for that purpose.

AGAIN, TEST VALIDITY AND RELIABILITY

A test must reflect only the competencies identified as critical in the job or task analysis—and as reflected in the performance objectives. This is why I advocate development of test items immediately following the writing of performance objectives—while the conditions, behaviors, and standards (or degrees of proficiency) are fresh in your mind. Make sure you have a test item for each critical aspect of the job to be measured. Once you have arranged all of your items into a complete test, you should have subject matter experts review it for content accuracy; this is phase III of the test development process. You will pilot the test using actual job incumbents. Their feedback will give you an idea of the test's ability to measure actual on-the-job competency requirements.

Several factors influence a test's reliability. The most important is the appropriateness and technical accuracy of the test items. They must evaluate realistic and practical aspects of the job performance. The number of test items is also important; while there must be at least one item for each

MT MechTech, Inc. APPRENTICE COMPETENCY PROFILE

Uniting Industry and Education for Metalworking Apprenticeships

Apprentice ____________ Company ____________ Date ____________

( ) Completed Rotation ( ) Transfer ( ) Lack of Work
( ) Cost Not Equal To Value ( ) Unacceptable Performance

Please evaluate this student regarding:

WORK HABITS AND ATTITUDES

Attention
( ) Always pays attention
( ) Almost always pays attention
( ) Generally pays attention
( ) Rarely pays attention
( ) Never pays attention

Ambition
( ) Very ambitious
( ) Usually energetic
( ) Average ambition
( ) Sporadic energy
( ) Lazy and little ambition

Responsibility
( ) Seeks responsibility
( ) Accepts willingly
( ) Usually cooperative
( ) Evades responsibility
( ) Refuses responsibility

Theory Knowledge
( ) Superior
( ) Above average
( ) Average
( ) Some
( ) Little

Cooperation with Others
( ) Always cooperative
( ) Above average
( ) Usually cooperative
( ) Limited cooperation
( ) Makes trouble

Attitude towards Supervision
( ) Always helpful
( ) Usually helpful
( ) Average respect
( ) Openly disrespectful

Emotional Adjustment
( ) Very stable adjustment
( ) Good adjustment
( ) Adequate adjustment
( ) Poorly adjusted
( ) Very unstable, moody

Interest in Trade
( ) Very high
( ) High
( ) Moderate
( ) A little
( ) Very little

Progress
( ) Outstanding
( ) Good
( ) Average
( ) Poor
( ) Unsatisfactory

Quality of Work
( ) Exceptional
( ) Above Average
( ) Average
( ) Poor
( ) Unsatisfactory

Care of Materials & Equipment
( ) Very careful
( ) Careful
( ) Generally careful
( ) Careless
( ) No regard

Potential for Success
( ) Highest
( ) Very good
( ) Probable
( ) Little
( ) None

SKILLS PERFORMANCE

E = EXCELLENT AA = ABOVE AVERAGE A = AVERAGE BA = BELOW AVERAGE

Accuracy	Speed	Skill/Machine	Accuracy	Speed	Skill/Machine
____	____	Drilling	____	____	Milling
____	____	Lathe work	____	____	Grinding
____	____	Sawing	____	____	Assembly
____	____	Production	____	____	Quality Assurance/Inspection
____	____	Numerical Control	____	____	Comprehension of Instruction
____	____	Safety	____	____	Performance of Assignments
____	____	Math Ability	____	____	____________
____	____	____________	____	____	____________

Would you accept this student back into your company at a later date? ( ) YES ( ) NO

Comments: ________________________________

Figure 9.2 Apprentice competency profile. (Reprinted with permission from MechTech, Inc.)

performance objective (including each enabling objective) in the training plan for a particular job, the more test items you include in the test, the better the test will be able to measure true competence. This should not be deemed a license to overkill; however, a four-step procedure doesn't warrant a 100-question test. Be sure you adequately and completely test all critical skills and knowledge.

Also, the order in which you present the items can affect test reliability. Just as improper distractors can cue the correct response, improper sequencing of items may give away correct behaviors or answers. Make sure item 1 does not tip off item 2. Also, make sure that the correct response to number 2 is not dependent on knowing the correct response to number 1. Each item must present an independent problem.

You may not be able to control some of the factors affecting reliability, but you should be aware of them. Try to create a consistent atmosphere for test taking; uniformity of the examination environment is important. The same test, administered once in a noisy and poorly lit factory area and then, again, in a quiet classroom will probably yield very different results. Also realize that cheating will skew test results. You should arrange for as much test security and exam proctoring as is necessary.

REFERENCES

Cantor, J. A. (1992). *Delivering Instruction to Adult Learners*. Toronto, Canada: Wall & Emerson, Inc.

Cantor, J. A. (1988, September). "How to Design, Develop, and Use Performance Tests," *Training & Development Journal*, 72–75.

Cantor, J. A. (1987, May). "Developing Multiple-Choice Test Items," *Training & Development Journal*, 85–88.

Resources and Organizations

ASSOCIATIONS

NATIONAL ASSOCIATION OF STATE AND TERRITORIAL APPRENTICESHIP DIRECTORS (NASTAD)

NASTAD was organized after World War II in order to promote a national apprenticeship system encompassing each state and territory that has laws and regulations governing apprenticeship. The goals of NASTAD are to work together to adopt uniform systems for continuous improvement of training and welfare of apprentices. Furthermore, NASTAD members work together to develop cooperative relationships with all employers and labor organizations and promote apprenticeship as an economic asset, and NASTAD cooperates with the U.S. Department of Labor, Bureau of Apprenticeship and Training to achieve a national apprenticeship system.

Each year, NASTAD elects an executive board. An annual meeting is held in a member state. Twenty-seven states, Washington, D.C., Guam, the Virgin Islands, and the Commonwealth of Puerto Rico are members. Further information can be obtained by contacting

Daniel F. Miles, Supervisor
Department of Labor and Industry
Job Service Division
Apprenticeship & Training Program
P.O. Box 1728
Helena, Montana 59624; (406) 444-3037

THE UNITED STATES APPRENTICESHIP ASSOCIATION (USAA)

The United States Apprenticeship Association is dedicated exclusively to the protection and betterment of apprenticeship. The USAA is the only private apprenticeship organization in the United States. Trade workers, union personnel, and management, as well as government and educational leaders, belong to the USAA. More information about USAA can be obtained from

Carl Horstrup, Chair
Industrial Technology & Apprenticeship Programs
Lane Community College
4600 East 30th Avenue
Eugene, OR 97405-0640; (503) 747-4501, Ext. 2843

ROCKY MOUNTAIN STATES APPRENTICESHIP CONFERENCE

The Rocky Mountain States Apprenticeship Conference was organized in 1969 by a group of industry, labor, government, and education leaders who recognized the vital need for skilled craftsmen in the Rocky Mountain states in an expanding industrial economy. They saw the Conference as a means of stimulating interest in and promoting the idea of trained men and women through apprenticeship. The Conference is composed of representatives from the states of Idaho, Arizona, Colorado, Montana, New Mexico, Wyoming, and Nevada. A steering committee composed of one representative and one alternate from management and one from labor from each of the participating states participate in the Conference's decision making. The steering committee assumes responsibility for guiding the destiny of the Conference, controlling its finances, raising funds annually for support, and establishing overall policy. The steering committee meets once a year at the annual Conference. More information about RMAC can be obtained from

Doris Romanisko, Steering Committee Chair
Montana Operating Engineers and AGC
Joint Apprenticeship and Training Trust
3100 Canyon Ferry Road
East Helena, MT 59635; (406) 227-5600

EASTERN STATES APPRENTICESHIP CONFERENCE

Each of the regional associations is organized in much the same manner. The ESAC holds annual meetings and conferences for the benefit of the eastern states. A point of contact for more information on ESAC is

Walt Purzycki, Administrator
Apprenticeship and Training Section
Division of Employment and Training

Delaware Department of Labor
P.O. Box 9499
Newark, DE 19714-9499; (302) 368-6995

SOUTHERN STATES REGIONAL APPRENTICESHIP CONFERENCE

A point of contact for the Southern States Conference is

Julian Palmer
U.S. Department of Labor
Atlanta Regional Office
Atlanta, GA 30367; (404) 347-4405

GOVERNMENT ASSISTANCE

Federal and state departments of labor can offer assistance in getting an apprenticeship program under way. These officials can also help in bringing together the business and labor and educational organizations to effect a cooperative apprenticeship program. Appendix D is a listing of U.S. Department of Labor, Bureau of Apprenticeship and Training state offices. Appendix E is a listing of labor commissioners/regional directors state-level contacts.

NEWSLETTERS

The Sentinel is published by the United States Apprenticeship Association (see address above). *The Sentinel* keeps its readers abreast of current trends and issues relating to apprenticeship training. Topics include federal legislation, innovative programs, concerns, solutions, and helpful tips.

JOURNALS AND PERIODICALS

Technical & Skills Training is published by the American Society for Training & Development. This monthly magazine focuses on all aspects of technical skills training, including apprenticeship. It is a practical and instructor-oriented publication that often contains useful hints for new instructors, as well as experienced training managers.

BOOKS

A useful text for instructor development is *Delivering Instruction to Adult Learners* by Jeffrey A. Cantor, Wall & Emerson, Inc., Toronto, Canada;

(416) 467-8685. This book covers those topics for instructor development that were discussed in Chapter 9 on related instruction.

Other useful books include:

National Apprenticeship Program
by U.S. Department of Labor
Employment and Training Division
Washington, D.C.

Apprenticeship Past and Present
by U.S. Department of Labor
Employment and Training Division
Washington, D.C.

Apprenticeship
by U.S. Department of Labor
Employment and Training Division
Washington, D.C.

Work-Based Learning: Training America's Workers
by U.S. Department of Labor
Employment and Training Division
Washington, D.C.

Apprenticeship 2000: Short Term Research Projects
by U.S. Department of Labor
Employment and Training Division
Washington, D.C.

TEXTBOX 10.1

U.S. Department of Labor, Bureau of Apprenticeship and Training State Offices

Alabama
Vacant, State Director
USDL-BAT
Medical Forum Bldg., Room 648
950 22nd Street North
Birmingham, Alabama 35203
205-731-1308

Alaska
Gary Berg, State Director
USDL-BAT
Calais Building, Suite 201
3301 C Street
Anchorage, Alaska 99503
907-271-5035

Arizona
George F. Jones, State Director
USDL-BAT
Suite 302
3221 North 16th Street
Phoenix, Arizona 85016
602-640-2964

Arkansas
Kenneth W. Lamkin, State Director
USDL-BAT
Federal Building, Room 3507
700 West Capitol Street
Little Rock, Arkansas 72201
501-324-5415

California
Jerry G. Tabaracci, State Director
USDL-BAT
Suite 1090-N
1301 Clay Street
Oakland, California 94612-5217
510-637-2951

Colorado
Betty A. J. Diaz, State Director
USDL-BAT
U.S. Custom House
721 19th Street, Room 469
Denver, Colorado 80202
303-844-4793

Connecticut
Phyllis H. Isreal, State Director
USDL-BAT
Federal Building
135 High Street, Room 367
Hartford, Connecticut 06103
203-240-4311

Delaware
George W. Elseroad, State Director
USDL-BAT
Lock Box 36—Federal Building
844 King Street
Wilmington, Delaware 19801
302-573-6113

Florida
George Belcher, State Director
USDL-BAT
City Centre Building, Suite 4110
227 North Bronough Street
Tallahassee, Florida 32301
904-942-8336

Georgia
Wallace R. Johnson, State Director
USDL-BAT
Room 418
1371 Peachtree Street, N.E.
Atlanta, Georgia 30367
404-347-4403

Hawaii
Sharon Gau, State Director
USDL-BAT
Room 5113
300 Ala Moana Boulevard
Honolulu, Hawaii 96850
808-541-2519

Idaho
Barbara Adolay, State Director
USDL-BAT
Suite 128
3050 North Lakeharbor Lane
Boise, Idaho 83703-6217
208-334-1013

Illinois
David Wyatt, State Director
USDL-BAT
Room 708
230 South Dearborn Street
Chicago, Illinois 60604
312-353-4690

Indiana
John M. Delgado, State Director
USDL-BAT
Federal Building and U.S. Courthouse
46 East Ohio Street, Room 414
Indianapolis, Indiana 46204
317-226-7592

Iowa
Michael G. Harcourt, State Director
USDL-BAT
Federal Building, Room 715
210 Walnut Street
Des Moines, Iowa 50309
515-284-4690

Kansas
Dolores E. Engel, State Director
USDL-BAT
Federal Building, Room 247
444 SE Quincy Street
Topeka, Kansas 66683-3571
913-295-2624

Kentucky
Vacancy, State Director
USDL-BAT
Federal Building, Room 187-J
600 Martin Luther King Place
Louisville, Kentucky 40202
502-582-5223

Louisiana
Ernest Goff, State Director
USDL-BAT
3535 South Sherwood Forest Boulevard
Afton Village Condo, Room 133
Baton Rouge, Louisiana 70816
504-389-0263

Maine
Richard Grandmaison, State Director
USDL-BAT
Federal Building
68 Sewall Street, Room 401
Augusta, Maine 04330
207-622-8235

Maryland
Larry E. Ward, State Director
USDL-BAT
Federal Building—Charles Center
31 Hopkins Plaza, Room 1028
Baltimore, Maryland 21201
410-962-2676

Massachusetts
John F. O'Shaughnessy, State Director
USDL-BAT
11th Floor
One Congress Street
Boston, Massachusetts 02114
617-565-2291

Michigan
Owen F. Mayer, State Director
USDL-BAT
Room 304
801 South Waverly
Lansing, Michigan 48917
517-377-1640

Minnesota
Joseph R. Meagher, State Director
USDL-BAT
Federal Building and U.S. Courthouse
316 Robert Street, Room 134
St. Paul, Minnesota 55101
612-290-3951

Mississippi
Vacant, State Director
USDL-BAT
Federal Building, Suite 410
100 West Capital Street
Jackson, Mississippi 39269
601-965-4346

Missouri
Robert D. Rains, State Director
USDL-BAT
Robert A. Young Federal Building
1222 Spruce, Room 9.102E
St. Louis, Missouri 63103
314-539-2522

Montana
John Gillespie, State Director
USDL-BAT
Federal Office Building
301 South Park Avenue
Room 396, Drawer #10055
Helena, Montana 59626-0055
406-449-5261

Nebraska
Charles Stoudamire, State Director
USDL-BAT
Room 801
106 South 15th Street
Omaha, Nebraska 68102
402-221-3281

Nevada
George E. Ramsey, State Director
USDL-BAT
Room 311
301 Stewart Avenue
Las Vegas, Nevada 89125
702-388-6396

New Hampshire
Jill Houser, State Director
USDL-BAT
143 North Main Street, Suite 205
Concord, New Hampshire 03301
603-255-1444

New Jersey
Dennis M. Fitzgerald, State Director
USDL-BAT
Parkway Towers Building E,
3rd Floor
485—Route #1, South
Iselin, New Jersey 08830
908-750-9191

New Mexico
Ruben Dominguez, State Director
USDL-BAT
505 Marquette, NW, Room 830
Albuquerque, New Mexico 87102
505-766-2398

New York
John Griffin, State Director
USDL-BAT
Federal Building, Room 809
No. Pearl & Clinton Avenue
Albany, New York 12202
518-431-4008

North Carolina
Curtis Stamper, State Director
USDL-BAT
Somerset Park, Suite 205
4407 Bland Road
Raleigh, North Carolina 27609
919-790-2801

North Dakota
Vacant, State Director
USDL-BAT
New Federal Building, Room 428
657 2nd Avenue, North
Fargo, North Dakota 58102
701-239-5415

Ohio
Mary Ann Dayspring, State Director
USDL-BAT
200 North High Street, Room 605
Columbus, Ohio 43215
614-469-7375

Oklahoma
Cynthia McLain, State Director
USDL-BAT
215 Dean A. McGee Ave.
Oklahoma City, Oklahoma 73102
405-528-1500

Oregon
Vacancy, State Director
USDL-BAT
Federal Building, Room 629
1220 SW 3rd Avenue
Portland, Oregon 97204
503-326-3157

Pennsylvania
Jeanette Elliott, State Director
USDL-BAT
Federal Building
228 Walnut Street, Room 773
Harrisburg, Pennsylvania 17108
717-782-3496

Rhode Island
Howard T. Carney, State Director
USDL-BAT
Federal Building
100 Hartford Avenue
Providence, Rhode Island 02909
401-528-5198

South Carolina
James H. Lee, State Director
USDL-BAT
Strom Thurmond Federal Building
1835 Assembly Street, Room 838
Columbia, South Carolina 29201
803-756-5547

South Dakota
William J. Kober, State Director
USDL-BAT
Oxbow I Building, Room 204
2400 West 49th Street
Sioux Falls, South Dakota 57102
605-330-4326

Tennessee
Merlin Taylor, State Director
USDL-BAT
Airport Executive Plaza
1321 Murfreesboro Road,
Suite 541
Nashville, Tennessee 37217-2648
615-781-5318

Texas
James Barrett, State Director
USDL-BAT
VA Building, Room 2102
2320 LaBranch Street
Houston, Texas 77004
713-750-1696

Utah
Ronald Matteucci, State Director
USDL-BAT
Administration Building,
Room 1051
1745 West 1700 South
Salt Lake City, Utah 84104-3839
801-975-3650

Vermont
Vacancy, State Director
USDL-BAT
Federal Building
11 Elmwood Avenue, Room 629
Burlington, Vermont 05401
802-951-6278

Virginia
James F. Walker, State Director
USDL-BAT
700 Centre, Suite 546
704 East Franklin
Richmond, Virginia 23240
804-771-2488

Washington
Ronald Johnson, State Director
USDL-BAT
Renton Plaza, Suite 100
1400 Talbot Road South
Renton, Washington 98055
206-277-5214

West Virginia
Dana C. Daugherty, State Director
USDL-BAT
Federal Building
550 Eagan Street, Room 303
Charleston, West Virginia 25301
304-347-5141

Wisconsin
Terrence V. Benewich,
State Director
USDL-BAT
Federal Center, Room 303
212 East Washington Avenue
Madison, Wisconsin 53703
608-264-5377

Wyoming
Vacancy, State Director
USDL-BAT
American National Bank Building
1912 Capitol Avenue, Room 508
Cheyenne, Wyoming 82001-3661
307-772-2448

Apprenticeship Programs: Labor Standards for Registration

TITLE 29—LABOR

SUBTITLE A—OFFICE OF THE SECRETARY OF LABOR

PART 29—LABOR STANDARDS FOR THE REGISTRATION OF APPRENTICESHIP PROGRAMS

POLICIES AND PROCEDURES

On Tuesday, October 19, 1976, the Department of Labor published in the *Federal Register* (41 FR 46148) proposed registration standards for apprenticeship programs. These standards, in the form of the addition of a new Part 29 to 29 CFR subtitle A, were promulgated pursuant to the authority of section 1 of the National Apprenticeship Act of 1937 (29 U.S.C. 50), Reorganization Plan No. 14 of 1950 (64 Stat. 1267; 3 CFR 1949–53 Comp., p. 1007), the Copeland Act (40 U.S.C. 276c), and 5 U.S.C. 301.

A revised version of the proposed standards was issued in 1975 and published at 40 FR 11340 (3-10-75). Comments to this initial proposed rulemaking were considered at length by the Federal Committee on Apprenticeship and by the Department of Labor. This process resulted in the issuance of the proposed rulemaking on October 19, 1976. The Department invited interested persons to submit written views and comments before November 22, 1976, concerning the proposal, and numerous responses were received. The Department has studied these comments carefully and several editorial and clarifying changes have been incorporated into the regulation. However, Part 29, which is published as final today, is basically the same as the proposal of October 19.

This document was prepared under the direction of Hugh C. Murphy,

Reprinted from *Federal Register,* February 18, 1977, 42(34) 10138–10144.

Administrator, Bureau of Apprenticeship and Training. For further information about this document, contact:

James P. Mitchell, Deputy Administrator
Bureau of Apprenticeship and Training
Employment and Training Administration
Room 5000, Patrick Henry Building
Washington, D.C. 20213
Telephone No. 202-376-6585

This new part sets out labor standards, policies and procedures relating to the registration, cancellation and deregistration of apprenticeship programs and of apprenticeship agreements by the Bureau of Apprenticeship and Training (BAT), the recognition of a State Apprenticeship Council or Agency (SAC) as the appropriate agency for registering local apprenticeship programs for certain Federal purposes and the derecognition of a SAC.

Those provisions which caused significant comment are as follows:

1. In § 29.2, Definitions, the definition of "Federal purposes" in paragraph (k) was unclear to several persons. The definition in this section is very broad. However, those Federal purposes which this part affects are described in § 29.3(a), which reads as follows: "Eligibility for various Federal purposes is conditioned upon a program's conformity with apprenticeship program standards published by the Secretary of Labor in this part. For a program to be determined by the Secretary of Labor as being in conformity with these published standards the program must be registered with the Bureau or registered with and/or approved by a State Apprenticeship Agency or Council recognized by the Bureau. Such determination by the Secretary is made only by such registration." Examples of such Federal purposes are the Davis-Bacon Act and the Service Contract Act.

2. In § 29.3, Eligibility and procedure for Bureau registration of a program, some persons read paragraph (h) as being applicable to "unilateral" programs (i.e., to programs sponsored by employers not having a collective bargaining agreement with a union). The text makes it quite clear that paragraph (h) applies only to those potential sponsors who are parties to an existing collective bargaining agreement and then only in very limited circumstances. Paragraph (i) underscores this point; it states that where an employer or group of employers wishes to register an apprenticeship program and there is no existing collective bargaining agreement, the employer or group of employers are not required to deal with a union.

3. In § 29.4, Criteria for apprenticeable occupation, paragraph (c) states that an apprenticeable occupation "involves manual, mechanical or technical skills and knowledge which require a minimum of 2,000 hours of on-the-job work experience." Several persons had the impression that the Bureau of

Apprenticeship and Training would allow almost any presently-recognized apprenticeable occupation to be registered as long as it met a minimum standard of 2,000 hours of on-the-job experience. This is not the intent of the Bureau of Apprenticeship and Training, nor does the paragraph when read in connection with the rest of this part—particularly § 29.5, Standards of apprenticeship—allow such an interpretation. Although the Bureau of Apprenticeship and Training has recognized only a handful of occupations having a minimum requirement of 2,000 hours of on-the-job experience, as well as related instruction to supplement this work experience, the Department believes other such occupations may exist. By setting 2,000 hours of on-the-job work experience as the minimum criterion, the Department feels it will be better able to fulfill its responsibility under the Fitzgerald Act to promote apprenticeship.

4. In § 29.5, Standards of apprenticeship, a number of changes have been made.

Paragraph (b) (4) has been changed to emphasize that plans of self-study will not be automatically approved. Rather, each such proposed plan will be considered on its merits by the Bureau of Apprenticeship and Training, as well as all other forms of related training, before approval is given to a program.

Paragraph (b) (7) has been amended to include safety as one of the factors to be weighed by the Bureau of Apprenticeship and Training when it considers the proposed ratio of apprentices to journeymen.

Paragraph (b) (10) has been revised as follows (omitted words are in brackets; added words are italicized): "The [required] minimum qualifications *required by a sponsor* for persons entering [an] *the* apprenticeship program, with an eligible starting age not less than 16 years;"

Paragraph (b) (14) has been revised by adding the words in italics: "Assurance of qualified training personnel *and adequate supervision on the job;*"

5. In § 29.12(a), Recognition of State agencies, the language of paragraph (a) has been revised to clarify the legal effect of the Secretary's recognition of a State Apprenticeship Council. Paragraph (a) now reads: "(a) The Secretary's recognition of a State Apprenticeship Agency or Council (SAC) gives the SAC the authority to determine whether an apprenticeship program conforms with the Secretary's published standards and the program is, therefore, eligible for those Federal purposes which require such a determination by the Secretary. Such recognition of a SAC shall be accorded by the Secretary upon submission and approval of the following:"

6. In § 29.12, several commenters objected to the language of paragraph (b) (8). This paragraph requires the SAC to "provide that apprenticeship programs and standards of employers and unions in other than the building and construction industry, which jointly form a sponsoring entity on a

multistate basis and are registered pursuant to all requirements of this part by any recognized State Apprenticeship Agency/Council or by the Bureau, shall be accorded registration or approval reciprocity by any other State Apprenticeship Agency/Council or office of the Bureau if such reciprocity is requested by the sponsoring entity."

This provision was approved without dissent by the Federal Committee on Apprenticeship on September 8, 1976. It was the intent of the Committee to simplify the problems experienced by a relatively few number of apprenticeship programs. None of these programs are in the construction occupations. Rather the paragraph applies to those programs which are operated by large, industrial companies such as General Motors, Ford, Alco, etc. in conjunction with the locals of several large international unions.

The national standards for these programs are developed by the national office of the joint apprenticeship committee of the industry, in conjunction with the national staff of the Bureau of Apprenticeship and Training. The Department of Labor approves and publishes these standards. The local joint apprenticeship committee ordinarily adopts the approved national pattern standards without change, except for such local matters as those involving wage rates and affirmative action goals. The local programs, which are administered jointly by the employer and the union, are situated in large plants with a relatively stable work force employed on a year-round basis. Hence, these programs differ from the typical construction employer who operates on a multistate basis.

The construction industry employs a mobile work force primarily in seasonal jobs. In construction programs, because of the seasonality of construction work, the apprentice's on-the-job training will usually be interrupted several times during the course of his/her apprenticeship and the supervision will be provided by several employers. In multistate operations, it may be necessary to provide related instruction at several places.

In the non-construction programs which this paragraph will affect, the typical apprentice will be employed year-round at the same site by the same employer during the entire term of his/her apprenticeship, and will receive on-the-job training and supervision from the same employer. Although related training may not be conducted at the worksite, it will ordinarily be conducted at the same location throughout the entire term of the individual's apprenticeship.

The Department believes it is reasonable to make a distinction between apprenticeship programs in the construction industry and those in other industries because of the differences mentioned above. These differences have an effect on what factors are necessary to insure a proper apprenticeship program in a particular craft.

The Department believes it is reasonable to draw a distinction between those multistate non-construction employers who conduct an apprenticeship program jointly with a union and those who conduct a unilateral apprenticeship program. The local programs, in practice, adopt the occupation's national pattern standards which have been developed by the occupation's national joint apprenticeship committee in cooperation with the national office of BAT and published by the Department.

The program is administered not by the employer alone but by the local joint apprenticeship committee (JAC) composed of both employer and union representatives. These two elements have both mutual and conflicting interests in assuring that the apprenticeship program is properly operated. The result of this tension of interests is more likely to result in a proper training program than would be the case in a program operated unilaterally.

Because of the stable year-round work force at the worksite, the journeymen are able to reach an informed opinion on the quality of the apprenticeship program. Each of the journeymen pays a percentage of his/her wage for the operation of the program. These circumstances increase the likelihood that complaints about deficiencies in the program, if not corrected by the JAC, will reach the registration agency which can take corrective action.

7. In § 29.12(c), language has been added to make clear that currently-recognized State Apprenticeship Agencies and State Apprenticeship Councils retain their recognition during the 120-day period after the effective date of this part, as well as during any extension period granted by the Administrator.

8. Several persons believed that the requirements contained throughout § 29.12 represent an unwarranted intrusion of Federal control into the operations of the SACs. The Department believes that this conclusion is not correct.

As far as the Department knows, the recognized SACs are already in substantial conformity with the minimum standards set forth in this section, with the exception of paragraphs (b)(8) and (b)(10), which have been addressed earlier. Where they are not, paragraph (c) affords the State a 120-day period within which to conform. An extension of time may be granted by the Administrator of the Bureau for good cause.

It does not seem to the Department that it will be an undue hardship for the SACs to conform to the minimal requirements set forth in this part or to provide to the Department the information required by § 29.12(a), since recognition by the Secretary has important economic effects (as in the operation of the Davis-Bacon Act and the Service Contract Act) and important effects in promoting and improving the apprenticeship system. For these reasons it seems reasonable to the Department that the Secretary have documentary evidence that a recognized State agency is conforming to the minimum standards set forth in this part.

Some persons have read § 29.12(a)(5) in a manner which does not appear justified by the text. It requires a SAC to submit to the Bureau "a description of policies and operating procedures which depart from or impose requirements in addition to those prescribed in this part." While the Bureau has the right to approve or disapprove such variations, the purpose of this provision is not to enable the Bureau to control SACs or to dictate policies and procedures. Rather, it allows the Secretary to be informed of the policies and procedures of the SACs to which the Secretary has accorded recognition. The Department can then make its own judgment on whether these policies and procedures conflict with the requirements of this part.

9. Finally, some persons expressed reservations about the hearing procedures that are outlined in these regulations, primarily in § 29.9. Specifically, hearings are called for in the following circumstances.

(a) The deregistration of Bureau-registered programs (§ 29.7);

(b) Denials of a State agency's application for Bureau recognition (§ 29.12); and

(c) Withdrawal of Bureau recognition of a State Apprenticeship Agency or Council (§ 29.13). These hearings are available to the aggrieved parties specified in the respective sections, when such aggrieved parties have taken the steps required to trigger their hearing rights.

The Department has adopted the hearing procedures used in this part for a number of reasons. First: The hearing provisions are sound from a standpoint of due process and conform to well-settled principles of administrative law. Section 29.9 allows for the appointment of an administrative law judge. Moreover, the hearing provides a forum where both sides, in an adversary setting, may present and defend evidence.

Second: the hearing provisions in this part are virtually identical to those of 29 CFR Part 30, relating to Equal Opportunity in Apprenticeship. The Department is not aware of any serious complaints about this procedure. It is anticipated that hearings under Part 29 will be infrequent. Under these circumstances, it does not seem feasible to establish a separate appeals mechanism.

Accordingly, Title 29 of the Code of Federal Regulations is amended, effective March 21, 1977 by adding the following new Part 29:

AUTHORITY: Sec. 1, 50 Stat. 664, as amended (29 U.S.C. 50; 40 U.S.C. 276c; 5 U.S.C. 301); Reorganization Plan No. 14 of 1950, 64 Stat. 1267 (5 U.S.C. App., p. 534).

§ 29.1 Purpose and scope.

(a) The National Apprenticeship Act of 1937, section 1 (29 U.S.C. 50), authorizes and directs the Secretary of Labor "to formulate and promote the furtherance of labor standards necessary to safeguard the welfare of apprentices, to extend the application of such standards by encouraging the inclusion thereof in contracts of apprenticeship, to bring together employers and labor for the formulation of programs of apprenticeship, to cooperate with State agencies engaged in the formulation and promotion of standards of apprenticeship, and to cooperate with the Office of Education under the Department of Health, Education, and Welfare * * *." Section 2 of the Act authorizes the Secretary of Labor to "publish information relating to existing and proposed labor standards of apprenticeship," and to "appoint national advisory committees * * *." (29 U.S.C. 50a).

(b) The purpose of this part is to set forth labor standards to safeguard the welfare of apprentices, and to extend the application of such standards by prescribing policies and procedures concerning the registration, for certain Federal purposes, of acceptable apprenticeship programs with the U.S. Department of Labor, Employment and Training Administration, Bureau of Apprenticeship and Training. These labor standards, policies and procedures cover the registration, cancellation and deregistration of apprenticeship programs and of apprenticeship agreements; the recognition of a State agency as the appropriate agency for registering local apprenticeship programs for certain Federal purposes; and matters relating thereto.

(c) For further information about this Part 29, contact: Deputy Administrator, Bureau of Apprenticeship and Training, Employment and Training Administration, Room 5000, Patrick Henry Building, Washington, D.C. 20213, Telephone number (202) 376-6585.

§ 29.2 Definitions.

As used in this part:

(a) "Department" shall mean the U.S. Department of Labor.

(b) "Secretary" shall mean the Secretary of Labor or any person specifically designated by him.

(c) "Bureau" shall mean the Bureau of Apprenticeship and Training, Employment and Training Administration.

(d) "Administrator" shall mean the Administrator of the Bureau of Apprenticeship and Training, or any person specifically designated by him.

(e) "Apprentice" shall mean a worker at least 16 years of age, except where a higher minimum age standard is otherwise fixed by law, who is employed to learn a skilled trade as defined in § 29.4 under standards of apprenticeship fulfilling the requirements of § 29.5.

(f) "Apprenticeship program" shall mean a plan containing all terms and conditions for the qualification, recruitment, selection, employment and training of apprentices, including such matters as the requirement for a written apprenticeship agreement.

(g) "Sponsor" shall mean any person, association, committee, or organization operating an apprenticeship program and in whose name the program is (or is to be) registered or approved.

(h) "Employer" shall mean any person or organization employing an apprenticeship whether or not such person or organization is a party to an apprenticeship agreement with the apprentice.

(i) "Apprenticeship committee" shall mean those persons designated by the sponsor to act for it in the administration of the program. A committee may be "joint," i.e., it is composed of an equal number of representatives of the employer(s) and of the employees represented by a bona fide collective bargaining agent(s) and has been established to conduct, operate, or administer an apprenticeship program and enter into apprenticeship agreements with apprentices. A committee may be "unilateral" or "non-joint" and shall mean a program sponsor in which a bona fide collective bargaining agent is not a participant.

(j) "Apprenticeship agreement" shall mean a written agreement between an apprentice and either his employer, or an apprenticeship committee acting as agent for employer(s), which agreement contains the terms and conditions of the employment and training of the apprentice.

(k) "Federal purposes" includes any Federal contract, grant, agreement or arrangement dealing with apprenticeship; and any Federal financial or other assistance, benefit, privilege, contribution, allowance, exemption, preference or right pertaining to apprenticeship.

(l) "Registration of an apprenticeship program" shall mean the acceptance and recording of such program by the Bureau of Apprenticeship and Training, or registration and/or approval by a recognized State Apprenticeship Agency, as meeting the basic standards and requirements of the Department for approval of such program for Federal purposes. Approval is evidenced by a Certificate of Registration or other written indicia.

(m) "Registration of an apprenticeship agreement" shall mean the acceptance and recording thereof by the Bureau or a recognized State Apprenticeship Agency as evidence of the participation of the apprentice in a particular registered apprenticeship program.

(n) "Certification" shall mean written approval by the Bureau of:

(1) A set of apprenticeship standards developed by a national committee or organization, joint or unilateral, for policy or guideline use by local affiliates, as substantially conforming to the standards of apprenticeship set forth in § 29.5; or

(2) An individual as eligible for probationary employment as an apprentice under a registered apprenticeship program.

(o) "Recognized State Apprenticeship Agency" or "recognized State Apprenticeship Council" shall mean an organization approved by the Bureau as an agency or council which has been properly constituted under an acceptable law or Executive order, and has been approved by the Bureau as the appropriate body for State registration and/or approval of local apprenticeship programs and agreements for Federal purposes.

(p) "State" shall mean any of the 50 States of the United States, the District of Columbia, or any territory or possession of the United States.

(q) "Related instruction" shall mean an organized and systematic form of instruction designed to provide the apprentice with knowledge of the theoretical and technical subjects related to his/her trade.

(r) "Cancellation" shall mean the termination of the registration or approval status of a program at the request of the sponsor or termination of an apprenticeship agreement at the request of the apprentice.

(s) "Registration agency" shall mean the Bureau or a recognized State Apprenticeship Agency.

§ 29.3 Eligibility and procedure for Bureau registration of a program.

(a) Eligibility for various Federal purposes is conditioned upon a program's conformity with apprenticeship program standards published by the Secretary of Labor in this part. For a program to be determined by the Secretary of Labor as being in conformity with these published standards the program must be registered with the Bureau or registered with and/or approved by a State Apprenticeship Agency or Council recognized by the Bureau. Such determination by the Secretary is made only by such registration.

(b) No apprenticeship program or agreement shall be eligible for Bureau registration unless (1) it is in conformity with the requirements of this part and the training is in an apprenticeable occupation having the characteristics set forth in § 29.4 herein, and (2) it is in conformity with the requirements of the Department's regulation on "Equal Employment Opportunity in Apprenticeship and Training" set forth in 29 CFR Part 30, as amended.

(c) Except as provided under paragraph (d) of this section, apprentices must be individually registered under a registered program. Such registration may be effected:

(1) By filing copies of each apprenticeship agreement; or

(2) Subject to prior Bureau approval, by filing a master copy of such agreement followed by a listing of the name and other required data, of each individual when apprenticed.

(d) The names of persons in their first 90 days of probationary employment as an apprentice under an apprenticeship program registered by the Bureau or a recognized State Apprenticeship Agency, if not individually registered under such program, shall be submitted immediately after employment to the Bureau or State Apprenticeship Agency for certification to establish the apprentice as eligible for such probationary employment.

(e) The appropriate registration office must be promptly notified of the cancellation, suspension, or termination of any apprenticeship agreement, with cause for same, and of apprenticeship completions.

(f) Operating apprenticeship programs when approved by the Bureau shall be accorded registration evidenced by a Certificate of Registration. Programs approved by recognized State Apprenticeship Agencies shall be accorded registration and/or approval evidenced by a similar certificate or other written indicia. When approved by the Bureau, national apprenticeship standards for policy or guideline use shall be accorded certification, evidenced by a certificate attesting to the Bureau's approval.

(g) Any modification(s) or change(s) to registered or certified programs shall be promptly submitted to the registration office and, if approved, shall be recorded and acknowledged as an amendment to such program.

(h) Under a program proposed for registration by an employer or employers' association, where the standards, collective bargaining agreement or other instrument, provide for participation by a union in any manner in the operation of the substantive matters of the apprenticeship program, and such participation is exercised, written acknowledgement of union agreement or "no objection" to the registration is required. Where no such participation is evidenced and practiced, the employer or employers' association shall simultaneously furnish to the union, if any, which is the collective bargaining agent of the employees to be trained, a copy of its application for registration and of the apprenticeship program. The registration agency shall provide a reasonable time period of not less than 30 days nor more than 60 days for receipt of union comments, if any, before final action on the approval.

(i) Where the employees to be trained have no collective bargaining agent, an apprenticeship program may be proposed for registration by an employer or group of employers.

§ 29.4 Criteria for apprenticeable occupations.

An apprenticeable occupation is a skilled trade which possesses all of the following characteristics:

(a) It is customarily learned in a practical way through a structured, systematic program of on-the-job supervised training.

(b) It is clearly identified and commonly recognized throughout an industry.

(c) It involves manual, mechanical or technical skills and knowledge which require a minimum of 2,000 hours of on-the-job work experience.

(d) It requires related instruction to supplement the on-the-job training.

§ 29.5 Standards of apprenticeship.

An apprenticeship program, to be eligible for registration/approval by a registration/approval agency, shall conform to the following standards:

(a) The program is an organized, written plan embodying the terms and conditions of employment, training, and supervision of one or more apprentices in the apprenticeable occupation, as defined in this part, and subscribed to by a sponsor who has undertaken to carry out the apprentice training program.

(b) The program standards contain the equal opportunity pledge prescribed in 29 CFR 30.3(b) and, when applicable, an affirmative action plan in accordance with 29 CFR 30.4, a selection method authorized in 29 CFR 30.5, or similar requirements expressed in a State Plan for Equal Employment Opportunity in Apprenticeship adopted pursuant to 29 CFR Part 30 and approved by the Department, and provisions concerning the following:

(1) The employment and training of the apprentice in a skilled trade;

(2) A term of apprenticeship, not less than 2,000 hours of work experience, consistent with training requirements as established by industry practice;

(3) An outline of the work processes in which the apprentice will receive supervised work experience and training on the job, and the allocation of the approximate time to be spent in each major process;

(4) Provision for organized, related and supplemental instruction in technical subjects related to the trade. A minimum of 144 hours for each year of apprenticeship is recommended. Such instruction may be given in a classroom through trade, industrial courses or by through trade, industrial or correspondence courses of equivalent value, or other forms of self-study approved by the registration/approval agency;

(5) A progressively increasing schedule of wages to be paid the apprentice consistent with the skill acquired. The entry wage shall be not less than the minimum wage prescribed by the Fair Labor Standards Act, where applicable, unless a higher wage is required by other applicable Federal law, State law, respective regulations, or by collective bargaining agreement;

(6) Periodic review and evaluation of the apprentice's progress in job performance and related instruction; and the maintenance of appropriate progress records;

(7) The numeric ratio of apprentices to journeymen consistent with proper supervision, training, safety, and continuity of employment, and applicable provisions in collective bargaining agreements, except where such ratios are expressly prohibited by the collective bargaining agreements. The ratio language shall be specific and clear as to application in terms of jobsite, work force, department or plant;

(8) A probationary period reasonable in relation to the full apprenticeship term, with full credit given for such period toward completion of apprenticeship;

(9) Adequate and safe equipment and facilities for training and supervision, and safety training for apprentices on the job and in related instruction;

(10) The minimum qualifications required by a sponsor for persons entering the apprenticeship program, with an eligible starting age not less than 16 years;

(11) The placement of an apprentice under a written apprenticeship agreement as required by the State apprenticeship law and regulation, or the Bureau where no such State law or regulation exists. The agreement shall directly, or by reference, incorporate the standards of the program as part of the agreement;

(12) The granting of advanced standing or credit for previously acquired experience, training, or skills for all applicants equally, with commensurate wages for any progression step so granted;

(13) Transfer of employer's training obligation when the employer is unable to fulfill his obligation under the apprenticeship agreement to another employer under the same program with consent of the apprentice and apprenticeship committee or program sponsor;

(14) Assurance of qualified training personnel and adequate supervision on the job;

(15) Recognition for successful completion of apprenticeship evidenced by an appropriate certificate;

(16) Identification of the registration agency;

(17) Provision for the registration, cancellation and deregistration of the program; and requirement for the prompt submission of any modification or amendment thereto;

(18) Provision for registration of apprenticeship agreements, modifica-

tions, and amendments; notice to the registration office of persons who have successfully completed apprenticeship programs; and notice of cancellations, suspensions and terminations of apprenticeship agreements and causes therefor;

(19) Authority for the termination of an apprenticeship agreement during the probationary period by either party without stated cause;

(20) A statement that the program will be conducted, operated and administered in conformity with applicable provision of 29 CFR Part 30, as amended, or a State EEO in apprenticeship plan adopted pursuant to 29 CFR Part 30 and approved by the Department;

(21) Name and address of the appropriate authority under the program to receive, process and make disposition of complaints;

(22) Recording and maintenance of all records concerning apprenticeship as may be required by the Bureau of recognized State Apprenticeship Agency and other applicable law.

§ 29.6 Apprenticeship agreement.

The apprenticeship agreement shall contain explicitly or by reference:

(a) Names and signatures of the contracting parties (apprentice, and the program sponsor or employer), and the signature of a parent or guardian if the apprentice is a minor.

(b) The date of birth of apprentice.

(c) Name and address of the program sponsor and registration agency.

(d) A statement of the trade or craft in which the apprentice is to be trained, and the beginning date and term (duration) of apprenticeship.

(e) A statement showing (1) the number of hours to be spent by the apprentice in work on the job, and (2) the number of hours to be spent in related and supplemental instruction which is recommended to be not less than 144 hours per year.

(f) A statement setting forth a schedule of the work processes in the trade or industry division in which the apprentice is to be trained and the approximate time to be spent at each process.

(g) A statement of the graduated scale of wages to be paid the apprentice and whether or not the required school time shall be compensated.

(h) Statements providing:

(1) For a specific period of probation during which the apprenticeship agreement may be terminated by either party to the agreement upon written notice to the registration agency;

(2) That, after the probationary period, the agreement may be cancelled at the request of the apprentice, or may be suspended, cancelled, or terminated by the sponsor, for good cause, with due notice to the apprentice and

a reasonable opportunity for corrective action, and with written notice to the apprentice and to the registration agency of the final action taken.

(i) A reference incorporating as part of the agreement the standards of the apprenticeship program as it exists on the date of the agreement and as it may be amended during the period of the agreement.

(j) A statement that the apprentice will be accorded equal opportunity in all phases of apprenticeship employment and training, without discrimination because of race, color, religion, national origin, or sex.

(k) Name and address of the appropriate authority, if any, designated under the program to receive, process and make disposition of controversies or differences arising out of the apprenticeship agreement when the controversies or differences cannot be adjusted locally or resolved in accordance with the established trade procedure or applicable collective bargaining provisions.

§ 29.7 Deregistration of Bureau-registered program.

Deregistration of a program may be effected upon the voluntary action of the sponsor by a request for cancellation of the registration, or upon reasonable cause, by the Bureau instituting formal deregistration proceedings in accordance with the provisions of this part.

(a) *Request by sponsor.* The registration officer may cancel the registration of an apprenticeship program by written acknowledgement of such request stating, but not limited to, the following matters:

(1) The registration is canceled at sponsor's request, and effective date thereof;

(2) That, within 15 days of the date of the acknowledgement, the sponsor shall notify all apprentices of such cancellation and the effective date; that such cancellation automatically deprives the apprentice of his/her individual registration; and that the deregistration of the program removes the apprentice from coverage for Federal purposes which require the Secretary of Labor's approval of an apprenticeship program.

(b) *Formal deregistration.*—(1) *Reasonable cause.* Deregistration proceedings may be undertaken when the apprenticeship program is not conducted, operated, and administered in accordance with the registered provisions or the requirements of this part, except that deregistration proceedings for violation of equal opportunity requirements shall be processed in accordance with the provisions under 29 CFR Pat 30, as amended;

(2) Where it appears the program is not being operated in accordance with the registered standards or with requirements of this part, the registration officer shall so notify the program sponsor in writing;

(3) The notice shall (i) be sent by registered or certified mail, with return receipt requested; (ii) state the shortcoming(s) and the remedy required; and (iii) state that a determination of reasonable cause for deregistration will be made unless corrective action is effected within 30 days;

(4) Upon request by the sponsor for good cause, the 30-day term may be extended for another 30 days. During the period for correction the sponsor shall be assisted in every reasonable way to achieve conformity;

(5) If the required correction is not effected within the allotted time, the registration officer shall send a notice to the sponsor, by registered or certified mail, return receipt requested, stating the following:

(i) The notice is sent pursuant to this subsection;

(ii) Certain deficiencies (stating them) were called to sponsor's attention and remedial measures requested, with dates of such occasions and letters; and that the sponsor has failed or refused to effect correction;

(iii) Based upon the stated deficiencies and failure of remedy, a determination of reasonable cause has been made and the program may be deregistered unless within 15 days of the receipt of this notice, the sponsor requests a hearing;

(iv) If a request for a hearing is not made, the entire matter will be submitted to the Administrator, BAT, for a decision on the record with respect to deregistration.

(6) If the sponsor has not requested a hearing, the registration officer shall transmit to the Administrator, BAT, a report containing all pertinent facts and circumstances concerning the nonconformity, including the findings and recommendation for deregistration, and copies of all relevant documents and records. Statements concerning interviews, meetings and conferences shall include the time, date, place, and persons present. The Administrator shall make a final order on the basis of the record before him.

(7) If the sponsor requests a hearing, the registration officer shall transmit to the Secretary, through the Administrator, a report containing all the data listed in paragraph (6) above. The Secretary shall convene a hearing in accordance with § 29.9; and shall make a final decision on the basis of the record before him including the proposed findings and recommended decision of the hearing officer.

(8) At his discretion, the Secretary may allow the sponsor a reasonable time to achieve voluntary corrective action. If the Secretary's decision is that the apprenticeship program is not operating in accordance with the registered provisions or requirements of this part, the apprenticeship program shall be deregistered. In each case in which reregistration is ordered, the Secretary shall make public notice of the order and shall notify the sponsor.

(9) Every order of deregistration shall contain a provision that the sponsor shall, within 15 days of the effective date of the order, notify all registered

apprentices of the deregistration of the program; the effective date thereof; that such cancellation automatically deprives the apprentice of his/her individual registration; and that the deregistration removes the apprentice from coverage for Federal purposes which require the Secretary of Labor's approval of an apprenticeship program.

§ 29.8 Reinstatement of program registration.

Any apprenticeship program deregistered pursuant to this part may be reinstated upon presentation of adequate evidence that the apprenticeship program is operating in accordance with this part. Such evidence shall be presented to the Administrator, BAT, if the sponsor had not requested a hearing, or to the Secretary, if an order of deregistration was entered pursuant to a hearing.

§ 29.9 Hearings.

(a) Within 10 days of his receipt of a request for a hearing, the Secretary shall designate a hearing officer. The hearing officer shall give reasonable notice of such hearing by registered mail, return receipt requested, to the appropriate sponsor. Such notice shall include (1) a reasonable time and place of hearing, (2) a statement of the provisions of this part pursuant to which the hearing is to be held, and (3) a concise statement of the matters pursuant to which the action forming the basis of the hearing is proposed to be taken.

(b) The hearing officer shall regulate the course of the hearing. Hearings shall be informally conducted. Every party shall have the right to counsel, and a fair opportunity to present his/her case, including such cross-examination as may be appropriate in the circumstances. Hearing officers shall make their proposed findings and recommended decisions to the Secretary upon the basis of the record before them.

§ 29.10 Limitations.

Nothing in this part or in any apprenticeship agreement shall operate to invalidate—

(a) Any apprenticeship provision in any collective bargaining agreement between employers and employees establishing higher apprenticeship standards; or

(b) Any special provisions for veterans, minority persons or females in the standards, apprentice qualifications or operation of the program, or in the apprenticeship agreement, which is not otherwise prohibited by law, Executive order, or authorized regulation.

§ 29.11 Complaints.

(a) This section is not applicable to any complaint concerning discrimination or other equal opportunity matters; all such complaints shall be submitted, processed and resolved in accordance with applicable provision in 29 CFR Part 30, as amended, or applicable provisions of a State Plan for Equal Employment Opportunities in Apprenticeship adopted pursuant to 29 CFR Part 30 and approved by the Department.

(b) Except for matters described in paragraph (a) of this section, any controversy or difference arising under an apprenticeship agreement which cannot be adjusted locally and which is not covered by a collective bargaining agreement, may be submitted by an apprentice, or his/her authorized representative, to the appropriate registration authority, either Federal or State, which has registered and/or approved the program in which the apprentice is enrolled, for review. Matters covered by a collective bargaining agreement are not subject to such review.

(c) The complaint, in writing and signed by the complainant, or authorized representative, shall be submitted within 60 days of the final local decision. It shall set forth the specific matter(s) complained of, together with all relevant facts and circumstances. Copies of all pertinent documents and correspondence shall accompany the complaint.

(d) The Bureau or recognized State Apprenticeship Agency, as appropriate, shall render an opinion within 90 days after receipt of the complaint, based upon such investigation of the matters submitted as may be found necessary, and the record before it. During the 90-day period, the Bureau or State agency shall make reasonable efforts to effect a satisfactory resolution between the parties involved. If so resolved, the parties shall be notified that the case is closed. Where an opinion is rendered, copies of same shall be sent to all interested parties.

(e) Nothing in this section shall be construed to require an apprentice to use the review procedure set forth in this section.

(f) A State Apprenticeship Agency may adopt a complaint review procedure differing in detail from that given in this section provided it is proposed and has been approved in the recognition of the State Apprenticeship Agency accorded by the Bureau.

§ 29.12 Recognition of State agencies.

(a) The Secretary's recognition of a State Apprenticeship Agency or Council (SAC) gives the SAC the authority to determine whether an apprenticeship program conforms with the Secretary's published standards and the program is, therefore, eligible for those Federal purposes which require such a determination by the Secretary. Such recognition of a SAC shall be accorded by the Secretary upon submission and approval of the following:

(1) An acceptable State apprenticeship law (or Executive order), and regulations adopted pursuant thereto;

(2) Acceptable composition of the State Apprenticeship Council (SAC);

(3) An acceptable State Plan for Equal Employment Opportunity in Apprenticeship;

(4) A description of the basic standards, criteria, and requirements for program registration and/or approval; and

(5) A description of policies and operating procedures which depart from or impose requirements in addition to those prescribed in this part.

(b) *Basic requirements.* Generally the basic requirements under the matters covered in paragraph (a) of this section shall be in conformity with applicable requirements as set forth in this part. Acceptable State provisions shall:

(1) Establish the apprenticeship agency in (i) the State Department of Labor, or (ii) in that agency of State government having jurisdiction of laws and regulations governing wages, hours, and working conditions, or (iii) that State agency presently recognized by the Bureau, with a State official empowered to direct the apprenticeship operation;

(2) Require that the State Apprenticeship Council be composed of persons familiar with apprenticeable occupations and an equal number of representatives of employer and of employee organizations and may include public members who shall not number in excess of the number named to represent either employer or employee organizations. Each representative so named shall have no vote except where such members have a vote according to the established practice of a presently recognized council. If the State official who directs the apprenticeship operation is a member of the council, provision may be made for the official to have a tie-breaking vote;

(3) Clearly delineate the respective powers and duties of the State official and of the council;

(4) Clearly designate the officer or body authorized to register and deregister apprenticeship programs and agreements;

(5) Establish policies and procedures to promote equality of opportunity in apprenticeship programs pursuant to a State Plan for Equal Employment Opportunity in Apprenticeship which adopts and implements the requirements of 29 CFR Part 30, as amended, and to require apprenticeship programs to operate in conformity with such State Plan and 29 CFR Part 30, as amended;

(6) Prescribe the contents of apprenticeship agreements;

(7) Limit the registration of apprenticeship programs to those providing training in "apprenticeable" occupations as defined in § 29.4;

(8) Provide that apprenticeship programs and standards of employers and

unions in other than the building and construction industry, which jointly form a sponsoring entity on a multistate basis and are registered pursuant to all requirements of this part by any recognized State Apprenticeship Agency/Council or by the Bureau, shall be accorded registration or approval reciprocity by any other State Apprenticeship Agency/Council or office of the Bureau if such reciprocity is requested by the sponsoring entity;

(9) Provide for the cancellation, deregistration and/or termination of approval of programs, and for temporary suspension, cancellation, deregistration and/or termination of approval of apprenticeship agreements; and

(10) Provide that under a program proposed for registration by an employer or employers' association, and where the standards, collective bargaining agreement or other instrument provides for participation by a union in any manner in the operation of the substantive matters of the apprenticeship program, and such participation is exercised, written acknowledgement of union agreement or "no objection" to the registration is required. Where no such participation is evidenced and practiced, the employer or employers' association shall simultaneously furnish to the union, if any, which is the collective bargaining agent of the employees to be trained, a copy of its application for registration and of the apprenticeship program. The State agency shall provide a reasonable time period of not less than 30 days nor more than 60 days for receipt of union comments, if any, before final action on the application for registration and/or approval.

(c) *Application for recognition.* A State Apprenticeship Agency/Council desiring recognition shall submit to the Administrator, BAT, the documentation specified in § 29.12(a) of this part. A currently recognized Agency/Council desiring continued recognition by the Bureau shall submit to the Administrator the documentation specified in § 29.12(a) of this part on or before July 18, 1977. An extension of time within which to comply with the requirements of this part may be granted by the Administrator for good cause upon written request by the State agency but the Administrator shall not extend the time for submission of the documentation required by § 29.12(a). The recognition of currently recognized Agencies/Councils shall continue until July 18, 1977 and during any extension period granted by the Administrator.

(d) *Appeal from denial of recognition.* The denial by the Administrator of a State agency's application for recognition under this part shall be in writing and shall set forth the reasons for the denial. The notice of denial shall be sent to the applicant by certified mail, return receipt requested. The applicant may appeal such a denial to the Secretary by mailing or otherwise furnishing to the Administrator within 30 days of receipt of the denial, a notice of appeal addressed to the Secretary and setting forth the following items:

(1) A statement that the applicant appeals to the Secretary to reverse the Administrator's decision to deny its application;

(2) The date of the Administrator's decision and the date the applicant received the decision;

(3) A summary of the reasons why the applicant believes that the Administrator's decision was incorrect;

(4) A copy of the application for recognition and subsequent modifications, if any;

(5) A copy of the Administrator's decision of denial. Within 10 days of receipt of a notice of appeal, the Secretary shall assign an Administrative Law Judge to conduct hearings and to recommend findings of fact and conclusions of law. The proceedings shall be informal, witnesses shall be sworn, and the parties shall have the right to counsel and of cross-examination.

The Administrative Law Judge shall submit the recommendations and conclusions, together with the entire record to the Secretary for final decision. The Secretary shall make his final decision in writing within 30 days of the Administrative Law Judge's submission. The Secretary may make a decision granting recognition conditional upon the performance of one or more actions by the applicant. In the event of such a conditional decision, recognition shall not be effective until the applicant has submitted to the Secretary evidence that the required actions have been performed and the Secretary has communicated to the applicant in writing that he is satisfied with the evidence submitted.

(e) *State apprenticeship programs.*

(1) An apprenticeship program submitted for registration with a State Apprenticeship Agency recognized by the Bureau shall, for Federal purposes, be in conformity with the State apprenticeship law, regulations, and with the State Plan for Equal Employment Opportunity in Apprenticeship as submitted to and approved by the Bureau pursuant to 29 CFR 30.15, as amended;

(2) In the event that a State Apprenticeship Agency is not recognized by the Bureau for Federal purposes, or that such recognition has been withdrawn, or if no State Apprenticeship Agency exists, registration with the Bureau may be requested. Such registration shall be granted if the program is conducted, administered and operated in accordance with the requirements of this part and the equal opportunity regulation in 29 CFR Part 30, as amended.

§ 29.13 Derecognition of State agencies.

The recognition for Federal purposes of a State Apprenticeship Agency or State Apprenticeship Council (hereinafter designated "respondent"), may be withdrawn for the failure to fulfill, or operate in conformity with, the requirements of this part. Derecognition proceedings for reasonable cause shall be instituted in accordance with the following:

(a) Derecognition proceedings for failure to adopt or properly enforce a State Plan for Equal Employment Opportunity in Apprenticeship shall be processed in accordance with the procedures prescribed in 29 CFR 30.15.

(b) For causes other than those under paragraph (a) above, the Bureau shall notify the respondent and appropriate State sponsors in writing by certified mail, with return receipt requested. The notice shall set forth the following:

(1) That reasonable cause exists to believe that the respondent has failed to fulfill or operate in conformity with the requirements of this part;

(2) The specific areas of nonconformity;

(3) The needed remedial measures; and

(4) That the Bureau proposes to withdraw recognition for Federal purposes unless corrective action is taken, or a hearing request mailed, within 30 days of the receipt of the notice.

(c) If, within the 30-day period, respondent:

(1) Complies with the requirements, the Bureau shall so notify the respondent and State sponsors, and the case shall be closed;

(2) Fails to comply or to request a hearing the Bureau shall decide whether recognition should be withdrawn. If the decision is in the affirmative, the Administrator shall forward all pertinent data to the Secretary, together with the findings and recommendation. The Secretary shall make the final decision, based upon the record before him.

(3) Requests a hearing, the Administrator shall forward the request to the Secretary, and the procedures under § 29.9 shall be followed, with notice thereof to the State apprenticeship sponsors.

(d) If the Secretary determines to withdraw recognition for Federal purposes, he shall notify the respondent and the State sponsors of such withdrawal and effect public notice of such withdrawal. The notice to the sponsors shall state that, 30 days after the date of the Secretary's order withdrawing recognition of the State agency, the Department shall cease to recognize for Federal purposes, each apprenticeship program registered with the State agency unless, within that time, the State sponsor requests registration with the Bureau. The Bureau may grant the request for registration contingent upon its finding that the State apprenticeship program is operating in accordance with the requirements of his part and of 29 CFR Part 30, as amended. The Bureau shall make a finding on this issue within 30 days of receipt of the request. If the finding is in the negative, the State sponsor shall be notified in writing that the contingent Bureau registration has been revoked. If the finding is in the affirmative, the State sponsor shall be notified in writing that the contingent Bureau registration is made permanent.

(e) If the sponsor fails to request Bureau registration, or upon a finding of noncompliance pursuant to a contingent Bureau registration, the written notice to such State sponsor shall further advise the recipient that any actions or benefits applicable to recognition "for Federal purposes" are no longer available to participants in its apprenticeship program.

(f) Such notice shall also direct the State sponsor to notify, within 15 days, all its registered apprentices of the withdrawal of recognition for Federal purposes; the effective date thereof; and that such withdrawal removes the apprentice from coverage under any Federal provision applicable to his/her individual registration under a program recognized or registered by the Secretary of Labor for Federal purposes.

(g) A State Apprenticeship Agency or Council whose recognition has been withdrawn pursuant to this part may have its recognition reinstated upon presentation of adequate evidence that it has fulfilled, and is operating in accordance with, the requirements of this part.

Signed at Washington, D.C. this 15th day of February, 1977.

RAY MARSHALL,
Secretary of Labor

[FR Doc. 77-5212 Filed 2-17-77; 8:45 am]

Apprenticeable Occupations: U.S. Department of Labor

Occupational Title	Term in Years
Accordian maker	4
Acoustical carpenter	4
Actor (amusement and recreation)	2
Air and hydronic balance technician	3
Air-conditioning mechanic (automotive services)	1
Air-conditioning installer, window	3
Aircraft mechanic, armament	4
Aircraft mechanic, electrical	4
Aircraft mechanic, plumb and hydraulics	4
Aircraft-armament mechanic (government services)	4
Aircraft-photograph-equipment mechanic	4
Airframe and power plant mechanic	4
Airplane coverer (aircraft)	4
Airplane inspector	3
Alarm operator (government services)	1
Alteration tailor	2
Ambulance attendant (EMT)	1
Animal trainer (amusement and recreation)	2
Architectural coatings finisher	3
Arson and bomb investigator	2
Artificial-glass-eye maker	5
Artificial-plastic-eye maker	5
Asphalt-paving machine operator	3
Assembler-installer, general	2
Assembler, aircraft, power	2
Assembler, aircraft, structures	4
Assembler, electromechanical	4
Assembler, metal building	2
Assembly technician	2
Assistant press operator	2
Audio operator	2
Audio-video repairer	2
Auger press operator, manual control	2
Automobile cooling system diagnostic technician	2
Automobile-maintenance equipment servicer	4
Automobile-radiator mechanic	2

Reprinted from *Occupational Outlook Quarterly,* Winter 1991/92, pp. 33–37.

Occupational Title	Term in Years
Automated equipment engineer-technician	4
Automatic-equipment technician (telephone and telegraph)	4
Automobile mechanic	4
Automobile tester (automotive services)	4
Automobile upholsterer	3
Automobile-body repairer	4
Automobile-repair-service estimator	4
Automotive-generator-and-starter repairer	2
Aviation safety equipment technician	4
Aviation support equipment repairer	4
Avionics technician	4
Baker (bakery products)	3
Baker (hotel and restaurant)	3
Baker, pizza (hotel and restaurant)	1
Bakery-machine mechanic	3
Bank-note designer	5
Barber	2
Bartender	1
Batch-and-furnace operator	4
Battery repairer	2
Beekeeper (agriculture and agricultural service)	4
Ben-day artist	6
Bench hand (jewelry)	2
Bindery worker	4
Bindery-machine setter	4
Biomedical equipment technician	4
Blacksmith	4
Blocker-and-cutter, contact lens	1
Boatbuilder, wood	4
Boiler operator (any industry)	4
Boilerhouse mechanic	3
Boilermaker fitter	4
Boilermaker I	4
Boilermaker II mechanic	3
Book binder	5
Bootmaker, hand	1
Bracelet and brooch maker	4
Brake repairer (automotive services)	2
Bricklayer (brick and tile)	4
Bricklayer, firebrick and refractory tile	4
Bricklayer (construction)	3
Brilliandeer-lopper (jewelry)	3
Butcher, all-round	3
Butcher, meat (hotel and restaurant)	3
Buttermaker (dairy products)	2
Cabinetmaker	4
Cable installer-repairer	3
Cable splicer	4
Cable television installer	1
Cable tester (telephone and telegraph)	4
Calibration laboratory technician	4
Camera operator	3
Camera repairer	2
Canal-equipment mechanic	2
Candy maker	3
Canvas worker	3
Car repairer (railroad locomotive and car building)	4

Occupational Title	Term in Years
Carburetor mechanic (automotive services)	4
Card cutter, jacquard	4
Card grinder (asbestos products)	4
Carpenter	4
Carpenter, maintenance	4
Carpenter, mold	6
Carpenter, piledriver	4
Carpenter, rough	4
Carpenter, ship (ship and boat building and repairing)	4
Carpet cutter (retail trade)	1
Carpet layer	3
Cartoonist, motion picture	3
Carver, hand	4
Cash-register servicer	3
Casing-in-line setter (printing and publishing)	4
Casket assembler	6
Caster (jewelry)	2
Caster (nonferrous metal alloys and primary products)	2
Cell maker (chemicals)	1
Cement mason	2
Central-office installer (telephone and telegraph)	4
Central-office repairer	4
Chaser (jewelry; silverware)	4
Cheesemaker	2
Chemical operator III	3
Chemical-engineering technician	4
Chemical-laboratory technician	4
Chief of party (professional and kindred)	4
Chief operator (chemicals)	3
Child care development specialist	2
Chimney repairer	1
Clarifying-plant operator (textiles)	1
Cloth designer	4
Coin-machine-service repairer	3
Colorist, photography	2
Commercial designer	4
Complaint inspector (light, heat, and power)	4
Composing-room machinist	6
Compositor	4
Computer programmer	2
Computer-peripheral-equipment-operator	1
Construction-equipment-mechanic	4
Contour wire specialist, denture	4
Conveyor-maintenance mechanic	2
Cook (any industry)	2
Cook (hotel and restaurant)	3
Cook, pastry (hotel and restaurant)	3
Cooling tower technician	2
Coppersmith (ship and boat building and repairing)	4
Coremaker	4
Cork insulator, refrigeration plant	4
Correction officer	1
Corrosion-control fitter	4
Cosmetologist	2
Counselor	2
Cupola tender	3
Custom tailor (garment)	4
Customer service representative	3

Occupational Title	Term in Years
Cutter, machine I	3
Cylinder grinder (printing and publishing)	5
Cylinder-press operator	4
Dairy equipment repairer	3
Dairy technologist	4
Decorator (any industry)	4
Decorator (glass manufacturing)	4
Dental assistant	1
Dental ceramist	2
Dental-equipment installer and servicer	3
Dental-laboratory technician	3
Design and patternmaker (boot and shoe)	2
Design drafter, electromechanisms	4
Detailer	4
Diamond selector (jewelry)	4
Dictating-transcribing-machine servicer	3
Die designer	4
Die finisher	4
Die maker (jewelry)	4
Die maker (paper goods)	4
Die maker, bench, stamping	4
Die maker, stamping	3
Die maker, trim	4
Die maker, wire drawing	3
Die polisher (nonferrous metal alloys and primary products)	1
Die setter (forging)	2
Die sinker	4
Diesel mechanic	4
Diesel-engine tester	4
Director, funeral	2
Director, television	2
Display designer (professional and kindred)	4
Displayer, merchandise	1
Door-closer mechanic	3
Dot etcher	5
Drafter, automotive design	4
Drafter, automotive design layout	4
Drafter, architectural	4
Drafter, cartographic	4
Drafter, civil	4
Drafter, commercial	4
Drafter, detail	4
Drafter, electrical	4
Drafter, electronic	4
Drafter, heating and ventilating	4
Drafter, landscape	4
Drafter, marine	4
Drafter, mechanical	4
Drafter, plumbing	4
Drafter, structural	3
Drafter, tool design	4
Dragline operator	1
Dredge operator (construction, mining)	1
Dressmaker	4
Drilling-machine operator	3
Dry cleaner	3
Dry-wall applicator	2
Electric-distribution checker	2
Electric-meter installer I	4
Electric-meter repairer	4
Electric-meter tester	4
Electric-motor assembler and tester	4
Electric-motor repairer	4
Electric-motor-and-generator assembler	2

Occupational Title	Term in Years
Electric-sign assembler	4
Electric-tool repairer	4
Electric-track-switch maintainer	4
Electrical technician	4
Electrical-appliance repairer	3
Electrical-appliance servicer	3
Electrical-instrument repairer	3
Electrician	4
Electrician (ship and boat building and repairing)	4
Electrician (water transportation)	4
Electrician, aircraft	4
Electrician, automotive	2
Electrician, locomotive	4
Electrician, maintenance	4
Electrician, powerhouse	4
Electrician, radio	4
Electrician, substation	3
Electromechanical technician	3
Electromedical-equipment repairer	2
Electronic prepress system operator	5
Electronic-organ technician	2
Electronic-production-line-maintenance mechanic	1
Electronic-sales-and-service technician	4
Electronics mechanic	4
Electronics technician	4
Electronics tester	3
Electronics utility worker	4
Electrotyper	5
Elevating-grader-operator	2
Elevator constructor	4
Elevator repairer	4
Embalmer (personal service)	2
Embosser	2

Occupational Title	Term in Years
Embossing-press operator	4
Emergency medical technician	3
Engine model maker	4
Engine repairer, service	4
Engine turner (jewelry)	2
Engine-lathe set-up operator	2
Engine-lathe set-up operator, tool	2
Engineering assistant, mechanical equipment	4
Engineering model maker	2
Engraver glass	2
Engraver I	5
Engraver, block (printing and publishing)	4
Engraver, hand, hard metal	4
Engraver, hand, soft metal	4
Engraver, machine	4
Engraver, pantograph I	4
Engraver, picture (printing and publishing)	10
Engraving press operator	3
Envelope-folding-machine adjuster	3
Equipment installer (telephone and telegraph)	4
Estimator and drafter	4
Etcher, hand (print and publishing)	5
Etcher, photoengraving	4
Experimental mechanic (motor and bicycles)	4
Experimental assembler	2
Exterminator, termite	2
Extruder operator plastics	4
Fabricator-assembler, metal products	4
Farm-equipment mechanic I	3

Occupational Title	Term in Years
Farm-equipment mechanic II	4
Farmer, general (agriculture and agricultural service)	4
Farmworker, general I	1
Fastener technologist	3
Field engineer (radio and television broadcasting)	4
Field service engineer	2
Film develop	3
Film laboratory technician	3
Film laboratory technician I	3
Film or videotape editor	4
Finisher, denture	1
Fire apparatus engineer	3
Fire captain	3
Fire engineer	1
Fire fighter	3
Fire fighter, crash, fire	1
Fire inspector	4
Fire medic	3
Fire-control mechanic	2
Firer, kiln (pottery and porcelain)	3
Fish and game warden (government services)	2
Fitter (machine shop)	2
Fitter I (any industry)	3
Fixture maker (lighting fixtures)	2
Floor layer	3
Floral designer	1
Floor-covering layer (railroad locomotive and car building)	3
Folding-machine operator	2
Forge-shop-machine repairer	3
Forging-press operator I	1
Form builder (construction)	2
Former, hand (any industry)	2
Forming-machine operator	4
Foundry metallurgist	4

Occupational Title	Term in Years
Four-slide-machine setter	2
Fourdrinier-machine tender	3
Freezer operator (dairy products)	1
Fretted-instrument repairer	3
Front-end mechanic	4
Fuel injection servicer	4
Fuel-system-maintenance-worker	2
Fur cutter (fur goods)	2
Fur designer (fur goods)	4
Fur finisher (fur goods)	2
Furnace installer	3
Furnace installer and repairer	4
Furnace operator	4
Furniture designer	4
Furniture finisher	3
Furniture upholsterer	4
Furrier (fur goods)	4
Gang sawyer, stone	2
Gas appliance servicer	3
Gas utility worker	3
Gas-engine repairer	4
Gas-main fitter	4
Gas-meter mechanic I	3
Gas-regulator repairer	3
Gauger (petroleum products)	2
Gear hobber set-up operator	4
Gear-cutting mach set-up operator	3
Gear-cutting mach set-up operator, tool	3
Gem cutter (jewelry)	3
Geodetic computer	2
Glass bender (signs)	4
Glass blower	3
Glass blower, laboratory apparatus	4

Occupational Title	Term in Years
Glass installer (automotive services)	2
Glass-blowing-lathe operator	4
Glazier	3
Glazier, stained glass	4
Grader (woodworking)	4
Graphic designer	1
Greenskeeper II	2
Grinder I (clocks, watches, and allied products)	4
Grinder operator, tool, precision	4
Grinder set-up operator, universal	4
Gunsmith	4
Harness maker	3
Harpsichord maker	2
Hat-block maker (woodwork)	3
Hazardous-waste material technician	2
Head sawyer	3
Health care sanitary technician	1
Heat treater I	4
Heat-transfer technician	4
Heating/air-conditioning installer and servicer	3
Heavy forger	4
Horse trainer	1
Horseshoer	2
Horticulturist	3
Housekeeper	1
Hydraulic-press servicer (ammunition)	2
Hydroelectric-machinery mechanic	3
Hydroelectric-station operator	3
Hydrometer calibrator	2
Illustrator (professional and kindred)	4
Industrial designer	4
Industrial engineering technician	4
Injection-molding-machine operator	1
Inspector, building	3
Inspector, electromechanical	4
Inspector, outside production	4
Inspector, precision	2
Inspector, quality assurance	3
Inspector, motor vehicles	2
Inspector, set-up and lay-out	4
Instrument repairer (any industry)	4
Instrument technician (light, heat, and power)	4
Instrument maker	4
Instrument maker and repairer	5
Instrument mechanic (any industry)	4
Instrumentation technician	4
Instrument mechanic, weapons system	4
Insulation worker	4
Interior designer	2
Investigator, private	1
Jacquard-loom weaver	4
Jacquard-plate maker	1
Jeweler	2
Jig builder wood box	2
Job printer	4
Joiner (ship and boat building and repairing)	4
Kiln operator (woodworking)	3
Knitter mechanic	4
Knitting-machine fixer	4
Laboratory assistant	3
Laboratory assistant metallurgical	2

Occupational Title	Term in Years
Laboratory technician	1
Laboratory tester	2
Landscape gardener	4
Landscape management technician	1
Landscape technician	2
Last-model maker	4
Lather	3
Laundry-machine mechanic	3
Lay-out technician	4
Lay-out worker (any industry)	4
Lead burner	4
Leather stamper	1
Legal secretary	1
Letterer (professional and kindred)	2
Licensed practical nurse	1
Light technician	4
Line erector	3
Line installer-repairer	4
Line maintainer	4
Line repairer	3
Liner (pottery and porcelain)	3
Linotype operator (printing and publishing)	5
Lithograph-press operator tin	4
Lithographic platemaker	4
Locksmith	4
Locomotive engineer	4
Loft worker (ship and boat building and repairing)	4
Logger, all-round	2
Logging-equipment mechanic	4
Logistics engineer	4
Loom fixer	3
Machine assembler	2
Machine builder	2
Machine erector	4
Machine fixer (carpet and rug)	4
Machine fixer (textile)	3
Machine operator I	1
Machine repairer, maintenance	4
Machine set-up operator, paper	4
Machine set-up operator	2
Machine setter	3
Machine setter	4
Machine setter (clocks, watches, and allied products)	4
Machine setter (woodwork)	4
Machine try-out setter	4
Machinist	4
Machinist, automotive	4
Machinist, experimental	4
Machinist, linotype	4
Machinist, marine engine	4
Machinist, motion-pic equipment	2
Machinist, outside (ship and boat building and repairing)	4
Machinist, wood	4
Mailer	4
Maintenance mechanic (any industry)	4
Maintenance mechanic (grain and feed milling)	2
Maintenance mechanic (petroleum products; construction)	4
Maintenance repairer, industrial	4
Maintenance machinist	4
Maintenance mechanic (compressed and liquified gases)	4
Maintenance mechanic, telephone	3
Maintenance repairer, building	2
Manager, food service	3
Manager, retail store	3
Marble finisher	2

Occupational Title	Term in Years
Marble setter	3
Marine-service technician	3
Material coordinator (clerical)	2
Materials engineer	5
Meat cutter	3
Mechanical-engineering technician	3
Mechanic, endless track vehicle	4
Mechanic, industrial truck	4
Mechanical-unit repairer	4
Medical secretary	1
Medical-laboratory technician	2
Metal fabricator	4
Metal model maker (automotive)	4
Meteorological equipment repairer	4
Meteorologist	3
Meter repairer (any industry)	3
Miller, wet process	3
Milling-machine set-up operator	2
Millwright	4
Mine-car repairer	2
Miner I (mining and quarry)	1
Mock-up builder (aircraft)	4
Model and mold maker (brick and tile)	2
Model and mold maker, plaster	4
Model builder (furniture)	2
Model maker (clocks, watches, and allied products)	4
Model maker (aircraft manufacturing)	4
Model maker II	4
Model maker pottery	2
Model maker (automobile manufacturing)	4
Model maker, firearms	4
Model maker, wood	4
Mold maker (pottery and porcelain)	3
Mold maker II (jewelry)	2
Mold maker (jewelry)	4
Mold maker, die-casting and plastic molding	4
Mold setter	1
Molder	4
Molder, pattern (foundry)	2
Monotype-keyboard operator	3
Monument setter (construction)	4
Mosaic worker	3
Motor-grade operator	3
Motorboat mechanic	3
Motorcycle repairer	3
Multi-operation-forming-machine setter	4
Multi-competent clinical assistant	2
Multi-operation-machine operator	3
Neon-sign servicer	5
Nondestructive tester	1
Numerical-control-machine operator	4
Nurse assistant	1
Office-machine servicer	3
Offset-press operator I	4
Oil-burner-servicer and installer	2
Oil-field equipment mechanic	2
Operating engineer	3
Operational test mechanic	3
Optical-instrument assembler	2
Optician	5
Optician (optical goods)	4
Optician-dispensing	2

Occupational Title	Term in Years
Optomechanical technician	4
Ordinance artificer (government services)	3
Ornamental-iron worker	3
Ornamental-metal worker	4
Orthopedic-boot-and-shoe designer and maker	5
Orthotics technician	1
Orthotist	5
Orthodontic technician	2
Outboard-motor mechanic	2
Overhauler (textile)	2
Painter	3
Painter (professional and kindred)	1
Painter, hand (any industry)	3
Painter, shipyard (ship and boat building and repairing)	3
Painter, sign	4
Painter, transportation equipment	3
Pantograph-machinc sct-up operator	2
Paperhanger	2
Paralegal	3
Paramedic	2
Paste-up artist	3
Patternmaker (textiles)	3
Patternmaker (metal prod)	4
Patternmaker (stonework)	4
Patternmaker, all-around	5
Patternmaker, metal	5
Patternmaker, metal, bench	5
Patternmaker, plaster	3
Patternmaker, plastics	3
Patternmaker, wood	5
Pewter caster	3
Pewter fabricator	4
Pewter finisher	2
Pewter turner	4
Pewterer	2
Pharmacist assistant	1
Photo-equipment technician	3
Photocomposing-perforator-machine operator	2
Photoengraver	5
Photoengraving finisher	5
Photoengraving printer	5
Photoengraving proofer	5
Photogrammetric technician	3
Photographer retoucher	3
Photographer, lithographic	5
Photographer, photoengraving	6
Photographer, still	3
Photographic-equipment-maintenance technician	3
Photographic-plate maker	4
Piano technician	4
Piano tuner	3
Pilot, ship	1.5
Pinsetter adjuster, automatic	3
Pinsetter mechanic, automatic	2
Pipe coverer and insulator (ship and boat building)	4
Pipe fitter (construction)	4
Pipe organ builder	3
Pipe fitter (ship and boat building and repairing)	4
Pipe-organ tuner and repairer	4
Plant operator	3
Plant operator, furnace process	4
Plaster-pattern caster	5
Plasterer	2
Plastic tool maker	4
Plastic-fixture builder	4
Plastics fabricator	2
Plate finisher (printing and publishing)	6
Platen-press operator	4
Plater	3
Plumber	4

Occupational Title	Term in Years
Pneumatic-tool repairer	4
Pneumatic-tube repairer	2
Podiatric assistant	2
Police officer I	2
Pony edger (sawmill)	2
Post-office clerk	2
Pottery-machine operator	3
Power-plant operator	4
Power-saw mechanic	3
Power-transformer repairer	4
Powerhouse mechanic	4
Precision assembler	3
Precision assembler, bench	2
Precision-lens grinder	4
Press operator, heavy duty	4
Printer, plastic	4
Printer-slotter operator	4
Process/shipping technician	4
Program assistant	3
Programmer, engineering and scientific	4
Project printer (photofinishing)	4
Proof-press operator	5
Proofsheet corrector (printing and publishing)	4
Prop maker (amusement and recreation)	4
Propulsion-motor-and-generator repairer	4
Prospecting driller (petroleum products)	2
Prosthetics technician	4
Prosthetist (personal protective and medical devices)	5
Protective-signal installer	4
Protective-signal repairer	3
Private-branch-exchange installer (telephone and telegraph)	4
Private-branch-exchange repairer	4
Pump erector (construction)	2
Pump servicer	3
Pumper-gauger	3
Purchasing agent	4
Purification operator II	4
Quality-control inspector	2
Quality-control technician	2
Radiation monitor	4
Radio mechanic (any industry)	3
Radio repairer (any industry)	4
Radio station operator	4
Radiographer	4
Recording engineer	2
Recovery operator (paper)	1
Recreational vehicle mechanic	4
Refinery operator	3
Refrigeration mechanic (any industry)	3
Refrigeration unit repairer	3
Reinforcing metal worker	3
Relay technician	2
Relay tester	4
Repairer I (chemical)	4
Repairer, handtools	3
Repairer, heavy	2
Repairer, welding equipment	2
Repairer, welding system and equipment	3
Reproduction technician	1
Research mechanic (aircraft)	4
Residential carpenter	2
Retoucher, photoengraving	5
Rigger	3
Rigger (ship and boat building and repairing)	2
Rocket-engine-component mechanic	4
Rocket-motor mechanic	4
Roll threader operator	1

Occupational Title	Term in Years	Occupational Title	Term in Years
Roller engraver, hand	2	Shop optician, surface room	4
Roofer	3	Shop optician, benchroom	4
Rotogravure-press operator	4	Shop tailor (garment)	4
Rubber tester (rubber goods)	4	Siderographer (printing and publishing)	4
Rubber-stamp maker	4	Sign erector I	4
Rubberizing mechanic	4	Signal maintainer (railroad locomotive and car building)	4
Rug cleaner, hand	1	Silk-screen cutter	3
Saddle maker (leather)	2	Silversmith II	3
Safe and vault service mechanic	4	Sketch maker I (printing and publishing)	5
Salesperson, parts	2	Small-engine mechanic	2
Sample maker, appliances	4	Soft-tile setter (construction)	3
Sample stitcher (garment)	4	Soil-conservation technician	3
Sandblaster, stone	3	Solderer (jewelry)	3
Saw filer (any industry)	4	Sound mixer	4
Saw maker (cutlery and tools)	3	Sound technician	3
Scale mechanic	4	Spinner, hand	3
Scanner operator	2	Spring coiling machine setter	4
Screen printer	2	Spring maker	4
Screw-machine operator, multiple spindle	4	Spring repairer, hand	4
Screw-machine operator, single spindle	3	Stage technician	3
Screw-machine set-up operator	4	Station installer and repairer	4
Screw-machine set-up operator, single spindle	3	Stationary engineer	4
Script supervisor (motion pictures)	1	Steam service inspector	4
Service mechanic (automobile manufacturing)	2	Steel-die printer	4
Service planner	4	Stencil cutter	2
Sewing-machine repairer	3	Stereotyper	6
Sheet metal worker	4	Stoker erector-and-service	4
Ship propeller finisher	3	Stone carver	3
Shipfitter (ship and boat building and repairing)	4	Stone polisher	3
Shipwright (ship and boat building and repairing)	4	Stone setter (jewelry)	4
Shoemaker, custom	3	Stone-lathe operator	3
		Stonecutter, hand	3
		Stonemason	3
		Stripper	5
		Stripper, lithographic II	4
		Structural-steel worker	3
		Substation operator	4

Occupational Title	Term in Years
Supercargo (water transportation)	2
Surface-plate finisher	2
Swimming-pool servicer	2
Switchboard operator (light, heat, and power)	3
Tank setter (petroleum products)	2
Tap-and-die maker technician	4
Tape-recorder repairer	4
Taper	2
Taxidermist (professional and kindred)	3
Technician, submarine cable equipment	2
Telecommunications technician	4
Telecommunicator	4
Telegraphic-typewriter operator	3
Television and radio repairer	4
Template maker	4
Template maker, extrusion die	4
Terrazzo finisher	2
Terrazzo worker	3
Test equipment mechanic	5
Test technician (professional and kindred)	5
Test-engine operator	2
Tester	3
Testing and regulating technician	4
Thermometer tester	1
Tile finisher	2
Tile setter	3
Tool builder	4
Tool design checker	4
Tool designer	4
Tool grinder I	3

Occupational Title	Term in Years
Tool maker	4
Tool maker, bench	4
Tool-and-die maker	4
Tool-grinder operator	4
Tool-machine set-up operator	3
Tractor mechanic	4
Transformer repairer	4
Transmission mechanic	2
Treatment-plant mechanic	3
Tree surgeon (agriculture and agricultural service)	3
Tree trimmer	2
Trouble locator, test desk	2
Truck driver, heavy	1
Truck-body builder	4
Truck-crane operator	3
Tumor registrar	2
Tune-up mechanic	2
Turbine operator	4
Turret-lathe set-up operator	4
Upholsterer	2
Violin maker, hand	4
Wallpaper printer I	4
Wardrobe supervisor	2
Waste-treatment operator	2
Wastewater-treatment-plant operator	2
Watch repairer	4
Water treatment-plant operator (waterworks)	3
Weather observer	2
Web-press operator	4
Welder, arc	4
Welder, combination	3
Welder-fitter	4
Welding technician	4
Welding-machine operator, arc	3

Occupational Title	Term in Years	Occupational Title	Term in Years
Well-drill operator (construction)	4	Wire sawyer (stonework)	2
Wildland fire fighter specialist	1	Wire weaver, cloth	4
Wind tunnel mechanic	4	Wirer (office machine)	2
Wind-instrument repairer	4	Wood-turning-lathe operator	1
Wine maker (vinous liquor)	2		
		X-ray equipment tester	2

Bureau of Apprenticeship and Training State Offices and State Apprenticeship Councils

Alabama
USDL-BAT
Berry Building, Suite 102
2017 Second Avenue North
Birmingham, Alabama 35203
(205) 731-1308

Alaska
USDL-BAT
Federal Building and Courthouse
222 West Seventh Street,
Room 554
Anchorage, Alaska 99513
(907) 271-5035

Arizona
USDL-BAT
Suite 302
3221 North 16th Street
Phoenix, Arizona 85016
(602) 640-2964

Apprenticeship Services
Arizona Department of Economic Security
438 West Adams Street
Phoenix, Arizona 85003
(602) 252-7771

Arkansas
USDL-BAT
Federal Building, Room 3507
700 West Capitol Street
Little Rock, Arkansas 72201
(501) 378-5415

California
USDL-BAT
Room 350
211 Main Street
San Francisco, California 94105
(415) 744-6581

Division of Apprenticeship Standards
395 Oyster Point Boulevard
Fifth Floor
San Francisco, California 94080
(415) 737-2700

Colorado
USDL-BAT
U.S. Custom House
721 19th Street, Room 480
Denver, Colorado 80202
(303) 844-4793

Reprinted from *Occupational Outloook Quarterly,* Winter 1991/92, pp. 39–40.

Connecticut
USDL-BAT
Federal Building
135 High Street, Room 367
Hartford, Connecticut 06103
(203) 240-4311

Office of Job Training and Skill Development
Connecticut Labor Department
200 Folly Brook Boulevard
Wethersfield, Connecticut 06109
(203) 566-4724

Delaware
USDL-BAT
Lock Box 36, Federal Building
844 King Street
Wilmington, Delaware 19801
(302) 573-6113

Apprenticeship and Training
Department of Labor
Division of Employment and Training
Sixth Floor, State Office Building
820 North French Street
Wilmington, Delaware 19801
(302) 571-1908

District of Columbia
District of Columbia Apprenticeship Council
500 C Street NW
Suite 241
Washington, D.C. 20001
(202) 639-1415

Florida
USDL-BAT
City Centre Building, Suite 5117
227 North Bronough Street
Tallahassee, Florida 32301
(904) 681-7161

Bureau of Apprenticeship
Division of Labor and Employment Sect
1320 Executive Center Drive
Atkins Building, Second Floor
Tallahassee, Florida 32301
(904) 488-8332

Georgia
USDL-BAT
Room 418
1371 Peachtree Street NE
Atlanta, Georgia 30367
(404) 347-4403

Hawaii
USDL-BAT
Room 5113
300 Ala Moana Boulevard
Honolulu, Hawaii 96850
(808) 541-2518

Apprenticeship Division
Department of Labor and Industry Relations
830 Punch Bowl Street
Honolulu, Hawaii 96813
(808) 548-2520

Idaho
USDL-BAT
Suite 128
3050 North Lakeharbor Lane
Boise, Idaho 83724
(208) 334-1013

Illinois
USDL-BAT
Room 758
230 South Dearborn Street
Chicago, Illinois 60604
(312) 353-4690

Indiana
USDL-BAT
Federal Building and Courthouse
46 East Ohio Street, Room 414
Indianapolis, Indiana 46204
(317) 269-7592

Iowa
USDL-BAT
Federal Building, Room 637
210 Walnut Street
Des Moines, Iowa 50309
(515) 284-4690

Kansas
USDL-BAT
Federal Building, Room 256
444 SE Quincy Street
Topeka, Kansas 66683
(913) 295-2624

Kansas State Apprenticeship Council
Department of Human Resources
401 SW Topeka Boulevard
Topeka, Kansas 66603-3182
(913) 296-3588

Kentucky
Apprenticeship and Training
Department of Labor
620 South Third Street
Louisville, Kentucky 40202
(502) 588-4466

USDL-BAT
Federal Building, Room 187-J
600 Federal Place
Louisville, Kentucky 40202
(502) 582-5223

Louisiana
USDL-BAT
U.S. Postal Building, Room 1323
701 Loyola Street
New Orleans, Louisiana 70113
(504) 589-6103

Apprenticeship and Training
Louisiana Department of Labor
Office of Labor
1001 North 23rd Street
Baton Rouge, Louisiana 70804-9094
(504) 342-7820

Maine
USDL-BAT
Federal Building
P.O. Box 917
68 Sewall Street, Room 408-D
Augusta, Maine 04330
(207) 622-8235

Apprenticeship Standards
Bureau of Labor Standards
State House Station #45
Augusta, Maine 04333
(207) 289-4307

Maryland
USDL-BAT
Federal Building Charles Center
31 Hopkins Plaza, Room 1028
Baltimore, Maryland 21201
(301) 962-2676

Apprenticeship and Training
Department of Employment and Training
Room 213
1100 North Eutaw Street
Baltimore, Maryland 21201
(301) 333-5718

Massachusetts
USDL-BAT
11th Floor
One Congress Street
Boston, Massachusetts 02114
(617) 565-2291

Department of Labor and Industry
Division of Apprenticeship Training
Leverett Saltonstall Building
100 Cambridge Street
Boston, Massachusetts 02202
(617) 727-3488

Michigan
USDL-BAT
Room 304
801 South Waverly
Lansing, Michigan 48917
(517) 377-1746

Minnesota
USDL-BAT
Federal Building and U.S. Courthouse
316 Robert Street, Room 134
St. Paul, Minnesota 55101
(612) 290-3951

Division of Apprenticeship
Department of Labor and Industry
Space Center Building, Fourth Floor
443 Lafayette Road
St. Paul, Minnesota 55101
(612) 296-2371

Mississippi
USDL-BAT
Federal Building, Suite 1010
100 West Capital Street
Jackson, Mississippi 39269
(601) 965-4346

Missouri
USDL-BAT
1222 Spruce, Room 9.102E
St. Louis, Missouri 63103
(314) 539-2522

Montana
USDL-BAT
Federal Office Building
301 South Park Avenue
Room 394, Drawer-10055
Helena, Montana 59626-0055
(406) 449-5261

Apprenticeship and Training Bureau
Employment Policy Division
Department of Labor and Industry
P.O. Box 1728
Helena, Montana 59626-0055
(406) 444-4500

Nebraska
USDL-BAT
Room 801
106 South 15th Street
Omaha, Nebraska 68102
(402) 221-3281

Nevada
USDL-BAT
P.O. Building, Room 311
301 East Stewart Avenue
Las Vegas, Nevada 89101
(702) 388-6396

Nevada State Apprenticeship Council
505 East King Street, Room 601
Carson City, Nevada 89710
(702) 885-4850

New Hampshire
USDL-BAT
143 North Main Street
Concord, New Hampshire 03301
(603) 225-1444

New Hampshire Apprenticeship Council
19 Pillsbury Street
Concord, New Hampshire 03301
(603) 271-3176

New Jersey
USDL-BAT
Parkway Towers Building E
Third Floor
485 Route 1, South
Iselin, New Jersey 08830
(201) 750-9191

New Mexico
USDL-BAT
Room 16
320 Central Avenue SW
Albuquerque, New Mexico 87102
(505) 766-2398

Apprenticeship and Training
New Mexico Department of Labor
501 Mountain Road NE
Suite 106
Albuquerque, New Mexico 87102
(505) 841-8989

New York
USDL-BAT
Federal Building, Room 810
North Pearl and Clinton Avenues
Albany, New York 12202
(518) 472-4800

Employability Development
New York State Department of Labor
State Office Campus
Building #12, Room 140
Albany, NY 12240
(518) 457-6820

North Carolina
USDL-BAT
Somerset Park, Suite 205
4407 Bland Road
Raleigh, North Carolina 27609
(919) 790-2801

Apprenticeship and Training
North Carolina Department of Labor
Memorial Building
214 West Jones Street
Raleigh, North Carolina 27603
(919) 733-7533

North Dakota
USDL-BAT
New Federal Building, Room 428
653 Second Avenue North
Fargo, North Dakota 58102
(701) 239-5415

Ohio
USDL-BAT
Room 605
200 North High Street
Columbus, Ohio 43215
(614) 469-7375

Ohio State Apprenticeship Council
2323 West Fifth Avenue, Room 2140
Columbus, Ohio 43216
(614) 640-2242

Oklahoma
USDL-BAT
Alfred P. Murrah Federal Building
200 NW Fifth, Room 526
Oklahoma City, Oklahoma 73102
(405) 231-4814

Oregon
USDL-BAT
Federal Building, Room 526
1220 SW Third Avenue
Portland, Oregon 97204
(503) 221-3157

Apprenticeship and Training Division
Oregon Bureau of Labor and Industry
State Office Building—Room 32
800 N.E. Oregon Street
Portland, Oregon 97232
(503) 731-4072

Pennsylvania
State Director
USDL-BAT
Federal Building
228 Walnut Street, Room 773
Harrisburg, Pennsylvania 17108
(717) 782-3496

Apprenticeship and Training
Labor and Industry Building
7th and Forster Street, Room 1303
Harrisburg, Pennsylvania 17120
(717) 787-3687

Puerto Rico
Incentive to the Private Sector Program
Right to Employment Administration
P.O. Box 4452
505 Munoz Rivera Avenue
San Juan, Puerto Rico 00936
(809) 754-5181

Rhode Island
USDL-BAT
Federal Building
100 Hartford Avenue
Providence, Rhode Island 02909
(401) 273-7640

Apprenticeship and Training
Rhode Island State Apprenticeship
Shore Council
200 Elmwood Avenue
Providence, Rhode Island 02907
(401) 457-1858

South Carolina
USDL-BAT
S. Thurmond Federal Building
1835 Assembly Street, Room 838
Columbia, South Carolina 29201
(803) 765-5547

South Dakota
USDL-BAT
Courthouse Plaza, Room 107
300 North Dakota Avenue
Sioux Falls, South Dakota 57102
(605) 330-4326

Tennessee
USDL-BAT
Metroplex Business Park
460 Metroplex Drive, Suite 101-A
Nashville, Tennessee 37211
(615) 736-5408

Texas
USDL-BAT
VA Building, Room 2102
2320 LaBranch Street
Houston, Texas 77004
(713) 750-1696

Utah
USDL-BAT
Room 1051
1745 West 1700 South
Salt Lake City, Utah 84104
(801) 524-5700

Vermont
USDL-BAT
Burlington Square
96 College Street, Suite 103
Burlington, Vermont 05401
(802) 951-6278

Apprenticeship and Training
Department of Labor and Industry
120 State Street
Montpelier, Vermont 05602
(802) 828-2157

Virginia
USDL-BAT
Room 10-020
400 North Eighth Street
Richmond, Virginia 23240
(804) 771-2488

Apprenticeship and Training
Division of Labor and Industry
P.O. Box 12064
205 North Fourth Street,
Room M-3
Richmond, Virginia 23241
(804) 786-2381

Virgin Islands
Division of Apprenticeship and
Training
Department of Labor
P.O. Box 890, Christiansted
St. Croix, Virgin Islands 00802
(809) 773-1300

Washington
USDL-BAT
Room 950
1111 Third Avenue
Seattle, Washington 98101-3212
(206) 442-4756

Apprenticeship and Training
Department of Labor and Industry
ESAC Division
925 Plum Street
Olympia, Washington 98504-0631
(206) 753-3487

West Virginia
USDL-BAT
Federal Building
550 Eagan Street, Room 303
Charleston, West Virginia 25301
(304) 347-5141

Wyoming
USDL-BAT
J. C. O'Mahoney Center
2120 Capitol Avenue, Room 5013
P.O. Box 1126
Cheyenne, Wyoming 82001
(307) 772-2448

Wisconsin
USDL-BAT
Federal Center, Room 303
212 East Washington Avenue
Madison, Wisconsin 53703
(608) 264-5377

Bureau of Apprenticeship
Standards
Department of Industry, Labor
and Human Relations
7201 East Washington Avenue
Room 211-X
Madison, Wisconsin 53703
(608) 266-3133

Labor Commissioners'/Regional Directors' State-Level Contacts

Alabama
Mr. Jerry C. Ray
Commissioner
Department of Labor
Suite 620
100 N. Union
Montgomery, AL 36104
334/242-3460

Alaska
Mr. Tom Cashen
Commissioner
Department of Labor
P.O. Box 21149
1111 West 8th Street
Juneau, AK 99802-1149
907/465-2700

Arizona
Mr. Orlando Macias
Director
Department of Labor
800 W. Washington Street,
Room 102
P.O. Box 19070
Phoenix, AZ 85005-9070
602/542-4515

Arkansas
Mr. James L. Salkeld
Director
Department of Labor
10421 West Markham
Little Rock, AR 72205
501/682-4500

California
Mr. Lloyd W. Aubry, Jr.
Director
Department of Industrial
Relations
455 Golden Gate Avenue,
Room 4181
San Francisco, CA 94101
415/703-4590

Colorado
Mr. John J. Donlon
Executive Director
Department of Labor and
Employment
600 Grant Street, Room 900
Denver, CO 80203-3528
303/837-3800

Connecticut
Mr. James P. Butler
Commissioner
Department of Labor
200 Folly Brook Boulevard
Wethersfield, CT 06109
203/566-4384

Delaware
Mr. Darrell J. Minott
Secretary
Department of Labor
Carvel Office Building, 6th Floor
820 North French Street
Wilmington, DE 19801
302/577-2713

District of Columbia
Mr. Joseph P. Yeldell
Director
Department of Employment Services
500 C Street, NW, Suite 600
Washington, D.C. 20001
202/724-7112

Florida
Mr. Doug Jamerson
Secretary of Labor
Department of Labor and Employment Security
Hartman Building, Suite 300
2012 Capital Circle, NE
Tallahassee, FL 32399-2152
904/922-7021

Georgia
Mr. David Poythress
Commissioner
Department of Labor
Sussex Place, Suite 600
148 International Boulevard, NE
Atlanta, GA 30303-1751
404/656-3011

Guam
Mr. Juan Taijito
Director
Department of Labor
Government of Guam
P.O. Box 9970
Tamuning, GU 96931-9970
671/646-9241

Hawaii
Ms. Lorraine H. Akiba
Director
Department of Labor and Industrial Relations
830 Punchbowl Street, Room 321
Honolulu, HI 96813
808/586-8844

Idaho
Mr. Robert Purcell
Director
Department of Labor and Industrial Services
277 North 6th Street
Boise, ID 83720
208/334-3950

Illinois
Ms. Shinae Chun
Director
Department of Labor
160 N. LaSalle Street
Chicago, IL 60601
312/793-2800

Indiana
Mr. Kenneth J. Zeller
Commissioner
Department of Labor
402 W. Washington Street, Room W195
Indianapolis, IN 46204-2739
317/232-2663

Iowa
Mr. Allen J. Meier
Labor Commissioner
Division of Labor Services
1000 East Grand Street
Des Moines, IA 50319
515/281-3447

Kansas
Mr. Wayne Franklin
Secretary
Department of Human Resources
401 S.W. Topeka Boulevard
Topeka, KS 66603-3182
913/296-7474

Kentucky
Mr. Bill Riggs
Secretary of Labor
Labor Cabinet
1047 U.S. Highway, 127 South
Suite 4
Frankfort, KY 40601
502/564-3070

Louisiana
Ms. Gayle F. Truly
Secretary
Department of Labor
P.O. Box 94094
Baton Rouge, LA 70804-9094
504/342-3011

Maine
Ms. Valerie R. Landry
Commissioner
Department of Labor
20 Union Street
P.O. Box 309
State House Station #54
Augusta, ME 04332
207/287-3788

Maryland
Mr. John P. O'Connor
Commissioner
Department of Labor and Industry
501 St. Paul Place
Baltimore, MD 21202-2272
410/333-4179

Massachusetts
Ms. Christine E. Morris
Secretary of Labor
Executive Office of Labor
1 Ashburton Place, Room 2112
Boston, MA 02108
617/727-6573

Michigan
Mr. Lowell Perry
Director
Department of Labor
201 North Washington Square
P.O. Box 30015
Lansing, MI 48909
517/373-9600

Minnesota
Mr. Gary Bastian
Comissioner
Department of Labor and Industry
Space Center, 5th Floor
443 Lafayette Road
St. Paul, MN 55155
612/296-2342

Mississippi
Mr. Liston Thomasson
Executive Director
Employment Security Commission
1520 West Capitol Street
P.O. Box 1699
Jackson, MS 39215-1699
601/354-8711

Missouri
Mr. Christopher Kelly
Chairman
Labor and Industrial Relations
Commission
3315 W. Truman Boulevard
P.O. Box 599

Jefferson City, MO 65102
314/751-2461

Montana
Ms. Laurie Ekanger
Commissioner
Department of Labor and Industry
P.O. Box 1728
Helena, MT 59624
406/444-4487

Nebraska
Mr. Dan Dolan
Commissioner
Department of Labor
P.O. Box 94600
550 South 16th Street
Lincoln, NE 68509-4600
402/471-3405

Nevada
Mr. Frank T. MacDonald
Commissioner
1445 Hot Springs Road, Suite 108
Carson City, NV 89710
702/687-4850

New Hampshire
Ms. Diane Symonds
Acting Commissioner
Department of Labor
State Office Park South
P.O. Box 2076
95 Pleasant Street
Concord, NH 03301
603/271-3171

New Jersey
Mr. Peter J. Calderone
Commissioner
Department of Labor
Labor and Industry Building
John Fitch Plaza
P.O. Box CN 110
Trenton, NJ 08625-0110
609/292-2323

New Mexico
Mr. Clinton D. Harden
Secretary
Department of Labor
1401 Broadway
P.O. Box 1928
Albuquerque, NM 87103
505/841-8406

New York
Mr. John E. Sweeney
Commissioner
Department of Labor
State Campus, Building #12
Albany, NY 12240
518/457-2741

North Carolina
Mr. Harry E. Payne, Jr.
Commissioner
Department of Labor
4 West Edenton Street
Raleigh, NC 27601
919/733-7166

North Dakota
Mr. Craig Hagen
Commissioner
Department of Labor
State Capitol Building
600 East Boulevard
Bismarck, ND 58505
701/328-2660

Ohio
Mr. John P. Stozich
Director
Department of Industrial Relations
2323 West Fifth Avenue

P.O. Box 825
Columbus, OH 43126
614/644-2223

Oklahoma
Ms. Brenda Reneau
Commissioner
Department of Labor
4001 North Lincoln Boulevard
Oklahoma City, OK 73105-5212
405/528-1500 ext. 200

Oregon
Mr. Jack Roberts
Commissioner
Bureau of Labor and Industries
800 NE Oregon Street #32
Portland, OR 97232
503/731-4070

Pennsylvania
Mr. Johnny J. Butler
Secretary
Department of Labor and Industry
1700 Labor and Industry Building
7th and Forster Streets
Harrisburg, PA 17120
717/787-3157

Puerto Rico
Mr. Cesar Juan Almodovar Marchany
Secretary
Department of Labor and Human Resources
505 Munoz Rivera Avenue
G.P.O. Box 3088
Hato Rey, PR 00918
809/754-2119

Rhode Island
Ms. Edna S. Poulin
Director
Department of Labor
610 Manton Avenue
Providence, RI 02909
401/457-1869

South Carolina
Mr. Virgil W. Duffie, Jr.
Director
Department of Labor, Licensing and Regulations
3600 Forest Drive
P.O. Box 11329
Columbia, SC 29211
803/734-9594

South Dakota
Mr. Craig Johnson
Secretary
Department of Labor
700 Governors Drive
Pierre, SD 57501-2277
605/773-3103

Tennessee
Mr. Al Bodie
Commissioner
Department of Labor
Gateway Plaza, 2nd Floor
710 James Robinson Parkway
Nashville, TN 37243-0655
615/741-2582

Texas
Mr. Charles E. Haddock
Labor Commissioner
Department of Labor and Standards
920 Colorado Street
P.O. Box 12157
Austin, TX 78701
512/463-2829

Utah
Mr. Stephen Hadley
Chairman

Industrial Commission
160 East 300 South, 3rd Floor
Salt Lake City, UT 84144-6600

Vermont
Ms. Mary S. Hooper
Commissioner
Department of Labor and Industry
National Life Building
Drawer 20
Montpelier, VT 05620-3401
802/828-2288

Virginia
Mr. Theron J. "Skip" Bell
Commissioner
Department of Labor and Industry
13 South Thirteenth Street
Richmond, VA 23219
804/786-2377

Virgin Islands
Ms. Lisa Harris-Moorehead
Acting Commissioner
Department of Labor
2131 Hospital Street
Christiansted
St. Croix, U.S. VI 00820-4666
809/773-1994

Washington
Mr. Mark O. Brown
Director
Department of Labor and Industries
7273 Linderson Way, SW
Tumwater, WA 98504
206-956-4213

West Virginia
Ms. Shelby Leary
Commissioner
Division of Labor
State Capitol Complex, Bldg. 3
1800 Washington Street, East
Charleston, WV 25305
304/558-7890

Wisconsin
Ms. Carol Skornicka
Secretary
Department of Industry, Labor and Human Relations
201 East Washington Avenue
P.O. Box 7946
Madison, WI 53707-7946
608/266-7552

Wyoming
Mr. Frank S. Galeotos
Director
Department of Employment
Herschler Building, 2-East
122 West 25th Street
Cheyenne, WY 82002
307/777-7661

Bureau of Apprenticeship and Training Regional Offices

	Jurisdictions Served
Region I—Boston Ambrose "Red" Bittner Regional Director One Congress Street, 11th Floor Boston, MA 02114 617-565-2288	Connecticut, Maine, Massachusetts, New Hampshire, Rhode Island, Vermont
Region II—New York Albert Hudanish Regional Director Federal Building, Room 602 201 Varick Street New York, NY 10014 212-337-2313	New Jersey, New York, Puerto Rico, Virgin Islands
Region III—Philadelphia Joseph T. Hersh Regional Director Gateway Building, Room 13240 3535 Market Street Philadelphia, PA 19104 215-596-6417	Delaware, District of Columbia, Maryland, Pennsylvania, Virginia, West Virginia
Region IV—Atlanta Julian S. Palmer Regional Director Room 200 1371 Peachtree Street, NE Atlanta, GA 30367 404-347-4405	Alabama, Florida, Georgia, Kentucky, Mississippi, North Carolina, South Carolina, Tennessee

	Jurisdictions Served
Region V—Chicago Richard D. Swain Regional Director Room 708 230 South Dearborn Street Chicago, IL 60604 312-353-7205	Illinois, Indiana, Michigan, Minnesota, Ohio, Wisconsin
Region VI—Dallas Sally S. Hall Regional Director Federal Building, Room 628 525 Griffin Street Dallas, TX 75202 214-767-4993	Arkansas, Louisiana, New Mexico, Oklahoma, Texas
Region VII—Kansas City Isadore H. Gross, Jr. Regional Director 110 Main Street Room 1040 Kansas City, MO 64105-2112 816-426-3856	Iowa, Kansas, Missouri, Nebraska
Region VIII—Denver Carl K. Heninger Regional Director U.S. Custom House, Room 465 721 19th Street Denver, CO 80202 303-844-4791	Colorado, Montana, North Dakota, South Dakota, Utah, Wyoming
Region IX—San Francisco Vacant Regional Director Federal Building, Room 715 71 Stevenson Street San Francisco, CA 94105 415-744-6580	Arizona, California, Hawaii, Nevada

	Jurisdictions Served
Region X—Seattle William F. Wadsworth Regional Director 1111 Third Avenue, Room 925 Seattle, WA 98101-3212 206-553-5286	Alaska, Idaho, Oregon, Washington

Index